DOUGLAS DC-7

THEODORE GIANNA

THUNDERBOLT 47 PUBLICATIONS

email: aviator@thunderbolt47publications.com

AF823434

Front cover photo by Nordiska museet (Nordic Museum)
Back cover photo by Lars Söderström

Copyright © 2023 Theodore Gianna. (THUNDERBOLT 47 PUBLICATIONS)
This publication is produced under the jurisdiction of international copyright laws. All rights reserved. No part of this publication may be reproduced, distributed, or transmitted in any form or by any means, including photocopying, recording, electronic, mechanical methods, or otherwise, without the express permission of the publishers and copyright holders.
Every effort has been made to make this book's information as complete and accurate as possible. However, the Publisher and Author accept no liability for any errors nor be held responsible for any subsequent use of the published information, data, or specific details within this publication.

ISBN: 978-0-6456548-4-4 (Paperback)
First printing edition 2023

INTRODUCTION

The DC-7, manufactured by the most famous airliner manufacturer in the mid-twentieth century, the Douglas Aircraft Company, was the fastest piston-engine airliner to take to the sky. It ruled the skies for five short years, from its maiden flight on May 18, 1953, until manufacturing halted in 1958.

DC-7 production ceased in that year, not because the DC-7 was unpopular, but because it had arrived on the scene just one year before the Boeing 367-80 (Dash 80) made its first flight in 1954. That moment was the beginning of the end for the DC-7. The Dash 80 ushered in the arrival of the jet era. Airlines would now start talking about the new future of jet travel and the many opportunities that jet airliners would offer them. The DC-7 would face an uphill battle to attain success, unlike its predecessor, the DC-6, which was successful at every turn. The DC-7's fate was like an actor who tried hard to get a role in a Hollywood movie but was constantly knocked back by the studios.

Consequently, only 308 DC-7s were manufactured during the DC-7's five-year production period, compared to 704 for the DC-6. However, the DC-7 was an outstanding airliner. It was much faster than the DC-6B; it had a more spacious interior, was pleasant to fly, and was sleeker than any Douglas prop liner ever built. However, it carried the burden of being the final piston propeller airliner ever manufactured by the renowned Douglas Company.

This book provides the reader with a chronological account of the Douglas DC-7's production history, commencing in 1952 and ending in 1958. The events described involve both the Douglas Aircraft Company and the airlines that operated the DC-7.

As usual, I want to credit Pat Bukiri, who has again assisted me by allowing me to use the 'Midway Airport Historians' photos. For the people out there who are interested in classic airliners and airlines, you can't go past Pat and the 'Midway Airport Historians.' Here are the internet links. Be sure to check them out:

Flickr - https://flickr.com/photos/twa1049g/

Web page : http://www.midwayhistorians.com/photos.html

Facebook :
https://www.facebook.com/fly2midway/photos/a.630021567083169/3244901305595169/?type=3

Instagram : :
https://www.instagram.com/midwayhistorians/?utm_source=ig_embed&ig_rid=140a25b5-c51c-428b-9b80-93a459b8d387

ACRONYMS USED IN THE BOOK

B/U	BROKEN UP	LSD	LEASED	RET	RETURNED	SOC:	STRUCK OFF CHARGE
WFU	WITHDRAWN FROM USE	MGD	MERGED	RR	RE-REGISTERED	SCR	SCRAPPED
BT	BOUGHT	NN	NEW NAME	TFD	TRANSFERED	W/O	WRITTEN OFF
CVT	CONVERTED	NTU	NOT TAKEN UP	D/D	DATE DELIVERED	S/N	SERIAL NUMBER
FF	FIRST FLIGHT	REG	REGISTERED	C/N	CONSTRUCTION NUMBER	STD	STORED

THE DOUGLAS AIRPLANE COMPANY

The Douglas Aircraft Company was the world's most significant aircraft manufacturer in the 1950s. Its magnitude alone, whether measured in floor space of over 12 million square feet, employee count of over 65,000, or order backlog of two billion dollars in 1952, was a monumental indication of its success. The Douglas Corporation possessed the Midas touch. Everything it did was profitable.

Military production was the company's main focus and most profitable endeavor. Douglas extensively developed and tested numerous varieties of military aircraft, guided missiles, and high-speed research aircraft like the Douglas X-3 'Stiletto' for the armed forces.

However, its commercial aircraft division was also busy. In 1952 alone,11 domestic and 12 foreign airline operators ordered 364 DC-6 commercial airliners. In addition, Douglas received orders for 58 units of its most recent model, the DC-7 four-engine transport, which was designed to supersede the DC-6 series. The DC-7 first flew in 1953, but it was only conceived two years earlier.

1952

American Airlines' President C R Smith approached Douglas in 1951 to construct an extended version of the DC-6B to compete with TWA's introduction of the Super Constellation on US domestic trunk routes. The aircraft manufacturer, President Donald Douglas, his inner circle, and the Douglas design team were not enthusiastic. They doubted whether a market for such an aircraft and its design and construction would drain profit from the well-established DC-6 line. C R Smith would not give up; in January 1952, he offered to pay $40 million for the new DC-7s to supplement its fleet of 50 DC-6s, 6 DC-6As, and 25 DC-6Bs. That did the trick. Seeing as AAL was footing the majority of the bill, Douglas adopted the project, and the DC-7 would become a reality. Interestingly, United Airlines did not participate in the deliberations, as it did when AAL and UAL collaborated with Douglas to manufacture the DC-6. Thus, AAL became the launch customer for the DC-7.

DOUGLAS
PRE-ORDERS
AMERICAN AIRLINES

In January, American Airlines (AAL), the largest domestic carrier in the United States, ordered 25 DC-7s. The aircraft (an enlarged and more powerful version of the DC-6B) was to be equipped with four 3,250 horsepower Wright Turbo-Compound engines and have a cruising speed of 360 mph (579 kph). Initial delivery was expected to begin in late 1953.

In addition, the first DC-7 scheduled services were scheduled to commence on January 1, 1954. The possibility of a cruising speed of 360 mph meant that the DC-7 would reach well into the turboprop-transport speed range, with a fuel consumption significantly superior to aircraft equipped with full-turboprop or even piston-engine power plants of comparable output. The Wright compounded engine's specific consumption was as low as 0.39 lb. hp (0.195 kg).

The DC-7's permitted gross weight was increased by 16,800 lb. (7,620 kg) over the DC-6B, and its fuselage was slightly lengthened to accommodate 60/66 first-class passengers or 95 passengers in a 'coach' configuration. American Airlines paid nearly $1.60 million per aircraft.

NATIONAL AIRLINES

One of the smaller carriers ordered DC-7s before the other major carriers. In late February, National Airlines (NAL), the airline from Florida, signed a contract with Douglas Aircraft Company to purchase 4 Douglas DC-7s. This was the second order for DC-7s, powered by the Wright Turbo-Compound R-3350 engines rated at 3,250 horsepower each, providing a speed of 360 mph.

On March 17, NAL confirmed the order, which cost NAL $8 million, including spare parts. The first delivery was scheduled for the 4th quarter of 1953, with the others starting in January 1954.

The 68-passenger DC-7s would reduce the New York-Miami flight time from four to three and a half hours.

National's order was on top of 8 DC-6Bs and 14 Convairs already ordered, which would double its current fleet at a total cost of $30 million.

DELTA AIR LINES

The third carrier to place orders for DC-7 liners was Delta (DAL). The company's directors approved the purchase of 4 Douglas DC-7s at the cost of $7 million, including spares. Delivery was for early 1954. Delta's DC-7s would accommodate 69 passengers, and the airline claimed that the DC-7's cruising speed of 360 mph would make the Chicago-Miami nonstop service in just 3 hrs. 45 mins, or 30 mins sooner than its DC-6s.

UNITED AIRLINES

Five months after AAL placed an order for 25 DC-7s, United Airlines (UAL) finally did the same. UAL also ordered 25 units at a total cost of $ 40 million. Initial delivery was for early 1954. UAL's DC-7s would accommodate 62 passengers and have a lounge area. Moreover, it would carry some 10,000 lbs. (4,536 kg) of cargo. The aircraft's Gross takeoff weight would be 122,200 lbs. (55,338 kg). UAL's aircraft would also be powered by the Turbo-Compound engines (TCE). At the time, one TCE cost $64,000; United bought 140 of them, totaling $8,960,000!

THE WRIGHT R-3350 TURBO COMPOUND ENGINE

The Wright R-3350 Turbo Compound engine was initially created for the military. More than 80,000 flight hours had been logged by the engines in military service by 1952, and service testing was ongoing. The 3350 horsepower engine was also scheduled to power the Douglas DC-7s and Lockheed L-1049C Super Constellations for airline service in January 1954 and mid-1953, respectively.

Combining the benefits of both jet and piston engines, the TCE could produce more than one horsepower per cubic inch while using the least amount of fuel of any aircraft powerplant at the time. It permitted more dependable and cost-effective performance at the 'greater speed-payload-long-range' combination. When the DC-7 entered service, American Airlines was able to lower the flight time from Los Angeles to New York to 8 hrs. and 10 mins, saving almost one hour compared to current flight schedules.

The TCE, which included innovations such as a lifelong steel crankcase and fuel injection, gained greater power and efficiency by harnessing exhaust fumes without needing additional gasoline. Three tiny turbines mounted on the rear of the engine extracted energy from the exhaust velocity and transferred the power back to the propeller via a basic series of gears and fluid couplings. Power recovery from the exhaust added 20% extra power to the engine. The 20% increase in engine output could be converted to carry substantially larger payloads, more passengers, or attain an extended nonstop range, if necessary. Moreover, the turbine system worked autonomously, requiring no additional controls or supervision from the flight crew.

The Wright R-3350 TCE could actually produce up to 3,500 horsepower and weighed only 3,480 lb. (1578 kg). At 3,500 horsepower, the engine generated 1,500 more horsepower than the original Wright 18, which was first tested in 1936. Its diameter was 56.59 inches and length 91.80 inches. Specific fuel consumption ranged from 385 to 395 at cruising power.

When the DC-7 received the Wright R-3350 Turbo Compound engine, it could reach a top speed in excess of 400 mph (644 kph) and an average cruising speed of 360 mph (579 kph). The four Wright R-3350 turbo-compound engines developed a total of 13,000 horsepower during takeoff. The 3,250 horsepower of each engine represented an increase of approximately 30% over the most advanced Douglas DC-6B.

The Wright Turbo Cyclone 18R 3350 TC Radial Engine. (Smithsonian)

TITANIUM

Approximately 88% of the Douglas DC-7 engine nacelles' skin was made of titanium sheet stock. That was the first time titanium was used in commercial aircraft construction. The DC-7 substituted titanium for aluminum alloys and stainless steel in parts of the aft nacelles, some frames, the landing gear doors, and firewall webs. A weight saving of about 200 lb. (90.7 kg) per aircraft was achieved through the use of titanium, which has the strength of steel but weighs 56% less.

Comparing the differences to the DC-6B, on the DC-7's two outboard nacelles, an area of 11,800 square inches of .025 gage titanium weighed only 64 lb. (29 kg) compared with the 68 lb. (30.8 kg) of the 8,000 square inches of stainless steel in comparable nacelles of the DC-6B. The DC-7's inboard nacelles contained 11,300 square inches of titanium. In contrast, on the DC-6B, the 7,600 square inches of stainless steel weighed approximately the same. As for the DC-7's larger landing gear doors, Douglas utilized 2,000 square inches of titanium compared with 1,700 square inches of steel on the DC-6B's landing gear doors. A gauge for gauge substitution of titanium for steel in the 11,500 square inches of firewall resulted in a weight saving of around 64 lb. for the DC-7.

ENGINEERING & PRODUCTION

Early in October, Douglas Aircraft Company announced from Santa Monica that the engineering of the DC-7 Turbo-Compound transport was completed. On November 28 of the following month, Douglas started production on the DC-7 at Santa Monica. The year 1953 was the year of the DC-7.

1953

DC-7 ORDERS 1st QUARTER 1953

UNITS	UNIT COST	FLEET COST	SEAT MILES (PER YEAR)
58	$ 2,000,000	$ 116,000,000	3,423,160

On April 20, Douglas rolled out the first DC-7 in preparation for engine run-ups a few days later. The company made the standard changes that occur when a new aircraft goes on the flight line.

FIRST FLIGHT

The first flight of the Douglas DC-7 occurred on May 18 in Santa Monica, California. Since no prototype of this further development of the DC-4/6 series was constructed, the first DC-7 was considered a 'production prototype.'

The DC-7 took off at 11:38 am and was flown by Douglas chief test pilot John Martin. Martin flew the aircraft for 2 hrs. 12 mins, and when he was back on the ground, he stated that the DC-7 flew 'like a dream.' However, the plane was actually faster than had been estimated because, while the DC-7 was red-lined at 300 mph for the initial flight, Martin reported he had to keep cutting back on the power and pulling the nose up to keep the DC-7 from exceeding the precautionary limitation.

Douglas officials cited the minimal noise level of the aircraft's Wright R-3350 compounded engines as a significant source of satisfaction. When Martin flew the DC-7 over the field at 500 feet and entered a rapid climb at a maximum cruise power of 1,800 horsepower and 2,400 revolutions per minute, the decibel level from the ground was lessened. That was music to the ears of Douglas' representatives.

1. This is United Air Lines DC-7, N6315C, 'Mainliner Hawaii' in the late 1950s. With 55 DC-7 propliners, United Air Lines possessed the second greatest fleet. American Airlines was first with 58. (courtesy of Pat Bukiri and the Midway Airport Historians, via Tom Kalina)

DC-7

CHARACTERISTICS OF THE DC-7

Aerodynamically and structurally comparable to the DC-6, the DC-7 was just over 8 ft (2.44 m) longer at 108 ft. 11 inches (32.9 m. 0.2794 m). This extension accommodated up to 95 passengers in the 'coach' configuration and 69 passengers in the 'deluxe' layout. Despite a weight increase of 25,000 lbs. (11,340 kg) over the DC-6, the wing area and span remained the same at 1,463 sq. ft. (135.9 sq. m.) and 117 ft 6 inches. (35.81 m) In fact, the fundamental wing configuration (including the airfoil profile of the National Advisory Committee for Aeronautics, 'NACA' 23016-12) dated back to the 1939 design of the second series DC-4. The DC-7 exhibited the common characteristic shared by the Douglas family of airliners.

Regarding the airframe, a novel design element was using the main landing gear as a speed brake; thus, when the pilot encountered heavy turbulence or desired to make a steep descent, he could rapidly decelerate by exposing his landing gear from its housing at top speed. The landing gear consisted of two pairs of dual-type main wheels and one steerable nose wheel, which retracted forward.

The fuselage was divided into three main compartments, allowing for numerous passenger and cargo configurations. With the standard configuration, the available cargo capacity was 707 cubic ft. (20.02 cubic m) or 13,660 lbs. (6,191 kg). In the deluxe version, passenger comfort was well catered for with broad, reclining seats positioned adjacent to large, rectangular windows. Air conditioning at sea level was maintained up to a flight altitude of 12,500 ft. (3,810 m), then decreased
flight altitude of 12,500 ft. (3,810 m), then decreased
to the equivalent of 8,000 ft. (2,438 m) at 25,000 ft. (7620 m).

The domestic version of the DC-7 had a gross takeoff weight of 122,200 lb. and a maximum fuel load of 5,526 gallons. (20,918 l), which provided the DC-7 with an absolute range of 4,420 miles (7,113 km) at 15,000 ft. (4,572 m) or 3,900 miles (6,276 km) at 23,500 ft. (7,163 m).

The DC-7's stalling speed was 99 mph (159 kph), without flaps 127 mph (204 kph). Similar to the DC-6, the DC-7 also featured a double-slotted rear flap. Moreover, the design of the flap was aerodynamically and structurally sound.

The DC-7's mean gross weight was 105,000 lbs. (47,625 kg). In addition, it had a maximum speed of 404 mph (650 kph) at 22,000 feet (6,706 m) and a cruising speed of 358 mph (576 kph) at 24,000 ft. (7,315 m). This allowed the DC-7 to conduct transcontinental and transoceanic flights more quickly than any other commercial piston-engine aircraft in service.

With the DC-7's gross takeoff weight of 122.200 lb. (55,428 kg), the four- engine rate of climb was 1,145 ft./min. (348.996 m /min) at sea level, and the service ceiling was 23,900 ft. (7,285 m). At the mean gross weight of 105,000 lb. (47,628 kg), the four-engine service ceiling rose to 26,700 ft. (8,138 m). CAA field length at sea level and a takeoff weight of 122,200 lb. was quoted as 6,380 ft. (1,945 m), while the field length for a landing weight of 95,000 lb. (43,091 kg) was given as 5,510 ft. (1,679 m) at sea level.

DC-7 TRIALS

Preliminary DC-7 flight testing in early November revealed that the propliner exceeded several of its original performance standards. Due to Douglas' expedited testing program, early service with the world's fastest piston-powered aircraft was expected. For the first time in the company's history, the CAA certification tests were conducted using three airliners rather than a single prototype. This approach allowed a typical one-year curriculum to be compressed into four months. The following tests were completed: power plant development, landing and takeoff performance, propeller vibration, flying characteristics, and diving brake tests.

In addition, company pilots and test engineers determined the climb performance, stall speed, sound level, air conditioning, and cockpit emergency procedures of the DC-7. The entire testing regimen would not be completed until another 50 to 100 hours had passed. As a result, the CAA certification for the Douglas DC-7 was granted on November 30, 22 months after the first design drawings were completed. The three DC-7 test planes completed 204 flights during the CAA flight-testing program, which began on July 22. In late December, production of the Douglas DC-7 started at Santa Monica. The DC-7 was finally ready for operations.

AMERICAN AIRLINES

Five days before the inauguration of American Airlines' non-stop transcontinental service with DC-7s, one of the carrier's DC-7s set up a new airline record for the Los Angeles-Idlewild route by covering the 2,468 miles in 6 hrs. 31 mins; an average speed of around 380 mph assisted by tailwinds. It was a proving flight with 57 passengers onboard.

COMPETITIVE DC-7 TRANSCONTINENTAL FLIGHTS BY NAL, AAL

National Airlines and American Airlines' new Douglas DC-7s operated transcontinental services at record and near-record speeds on pre-inaugural flights during Thanksgiving week. National received its DC-7 at Santa Monica, California, and broke the Los Angeles-Miami transport speed record, flying the distance in 5 hrs. 50 mins and 13 secs. Eastern Air Lines had established the previous record in 1947 when one of its Constellations flew the route in 6 hrs. 25 mins.

Meanwhile, in two round-trip inaugural press flights, American made one eastbound hop in 6 hrs. 38 mins and the second in 6 hrs. 31 mins.

All flights were performed at altitudes from 21,000 to 25,000 ft (6,401 to 7,620 m). Tailwinds were officially defined as 'below average' for the time of year. NAL averaged 401 mph (645 kph), while AAL averaged 380 mph (611 kph) on one trip and 390 mph (627 kph) on the other. Speeds well above the averages were attained at various intervals throughout the journey.

FIRST TRANSCONTINENTAL SERVICE AND OTHER EVENTS

On November 29, American Airlines launched its first transcontinental DC-7 service, while National launched its maiden service on the New York-Miami route on December 15.

With the introduction of nonstop DC-7 service between New York and Los Angeles in late November, American Airlines began reducing hours from its transcontinental itineraries. Until December 20, when the second nonstop route was launched, American operated one daily nonstop flight on its 'Mercury' schedule.

The eastbound and westbound DC-7 services took 7 hrs. 15 mins and 7 hrs. 55 mins, respectively. That, together with the CAA's certification tests and nonstop transcontinental proving flights in November with full passenger loads, left no doubt that the DC-7s would comfortably complete all services.

During the proving flights, two of AA's DC-7s flew eastward in 6 hrs. 38 mins and 6 hrs. 31 mins, respectively, with the latter setting a new unofficial transcontinental record for an airliner with a full load. The quicker aircraft averaged 380 mph (612 kph) on the 2,474-mile (3,982 km) Great Circle route between Los Angeles and New York.

SPEED. The DC-7's key selling point was its speed. AA intended to capitalize on this speed by providing the fastest nonstop transcontinental flights between Boston, Philadelphia, Washington, Chicago, and Dallas.

In addition, American Airlines considered enhancing the DC-7s' cruising speed of 365 mph (587 kph). Essentially, each Wright R3350 Turbo Compound engine provided AA with 3,250 hp during takeoff and 1,800 hp in cruise mode. Subsequently, the airline intended to increase takeoff horsepower to 3,500 and cruise horsepower to 1,900. American's DC-7s was fitted with the enhanced Turbo Compound variant engine, which featured a substantial protective shield surrounding the turbine rotors. The DC-7's four Turbo Compounds burned 2,900 lb. (1,315 kg) of 115 octane fuel per hour.

OTHER ITEMS

- The DC-7's cabin noise level improved compared to the DC-6, which was about the same despite the increase of 2,400 hp in cruise flight and 3,000 hp during takeoff. The four-blade Hamilton Standard propellers also helped reduce the noise with their slower tip speeds.
- Cabin seats were more comfortable, particularly the foam rubber headrests.

- The cabin temperature was maintained at 8,000 ft. (2,440 m) while cruising at 25,000 ft. (7620 m). Standard DC-7 flight patterns were at 25,000 ft. altitude.
- The DC-7's main landing gear was strengthened. Subsequently, the landing gear could be extended at 300 mph. (483 kph).

2. American Airlines DC-7, N313AA, c/n 44134, 'Flagship Tennessee'. DEL: 1/16/1954. Five years later, sold to GALCO 2/25/1959. LSD to Overseas National Airways in April 1959, REG: N313A. BT by Charlotte Aircraft Corporation in 1963. WFU & B/U in 1964. Here it is seen at Chicago Midway Airport on a murky day in 1958. (courtesy of Pat Bukiri and the Midway Airport Historians)

3. American DC-7, 'Flagship Indiana', N317AA, c/n 44138, DEL: 2/13/1954. At Chicago Midway in 1956. Sold to GALCO 1/31/1959.LSD to Overseas National Airways, REG: N317A. Accident: (DBR) This DC-7 was on a flight from New York to Norfolk to pick up troops from NAS Chambers. The DC-7 had hydraulic issues and touched down 275 ft past the runway threshold. It went down the runway for 1,400 ft, before veering off to the left of the runway, hitting an embankment. No casualties. B/U. (courtesy of Pat Bukiri and the Midway Airport Historians)

In August, Delta-C & S Air Lines, which had already ordered 4 DC-7s in 1952, announced an order for 6 additional Douglas DC-7s, making it 10 in total (on May 1, 1953, Delta Air Lines and Chicago & Southern Air Lines merged). The total cost for the 6 DC-7s was $17.5 million.

Panagra (Pan American-Grace Airways) ordered 5 Douglas DC-7s in mid-October, becoming the fifth major airline to do so. Panagra's aircraft were the overseas version for service between New York and South American cities, including Buenos Aires, Argentina, Santiago, Chile, La Paz, Bolivia, Lima, Peru, Guayaquil, Ecuador, and the Panama Canal Zone. The Panagra order brought the total number of DC-7s on order to 69.

Then, in late November, Eastern Air Lines placed a $25m order for 12 Douglas DC-7s for delivery early in 1955. The order brought the number of DC-7s sold and ordered to 81. Eastern's fleet, one of the largest in the world in the 1950s, consisted of 11 DC-4s, 57 Martin 4-0-4s, 20 Constellations, and 14 Super Constellations.

The year's final order came from the 'World's Most Experienced Airline,' Pan American World Airways. In mid-December, Pan Am was the first to order 7 DC-7Bs, the overseas variant of the DC-7. Pan American's DC-7Bs were designed to transport 52 passengers and fly nonstop for 4,000 miles (6,437 km) with nearly 1,000 gal. (3,790 l) of fuel reserve. Pan Am's DC-7Bs could thus fly from New York to London in slightly more than 10 hours. The DC-7B fleet, to be delivered in 1955, cost the airline $14 million. The 'B' model had a maximum takeoff weight of 125,000 lb. (54432 kg) and a fuel capacity of 6,260 gallons (23,697 l).

DC-7 Domestic Specifications

POWERPLANTS	4 x WRIGHT 972TC18DA4 Turbo Compound Engines Takeoff: 3,250 hp / 2,900 rpm Normal : 2,700 hp / 2,600 rpm
CARGO CAPACITY lb. (kg)	12,740 (5,737)
MAX. SPEED mph (kph)	409 (658)
BEST CRUISE SPEED mph (kph)	367 (591)
CLIMB fpm (m/min)	448 (137)
SERVICE CEILING FT. (m)	27,900 (8,504)
TAKEOFF LENGTH ft. (m)	6,060 (1,847)
LANDING LENGTH ft (m)	5,840 (1,780)
MAX. STILL AIR RANGE mi. (km)	4,430 (7,129)
MAX. PAYLOAD RANGE mi. (km)	2,850 (4,586)
MAX. FUEL RANGE mi. (km)	4,340 (6,985)
EMPTY WEIGHT lb. (kg)	68,073 (30,877)
MTOW lb. (kg)	122,200 (55,429)
MAX. LANDING WEIGHT lb. (kg)	97,000 (43,998)
WINGSPAN ft (m)	117.6 (35.84)
LENGTH ft. (m)	108.11 (32.95)
HEIGHT ft. (m)	28.7 (8.74)
SEATING	69-99

NEW AIRLINERS

AIRCRAFT	NUMBER IN SCHEDULED SERVICE AT END OF 1952	NUMBER DELIVERED 1954-1955	TOTAL NUMBER OF NEW AIRCRAFT	TOTAL COST OF AIRCRAFT
DC-7	10	62	72	$144 MILLION
L-1049	31	60	91	$ 182 MILLION
DC-6B	106	26	132	$ 158.4 MILLION
CONVAIR CV-340	103	28	131	$ 91.7 MILLION
MARTIN 4-0-4	100	-	100	$ 70 MILLION
TOTAL	348	176	526	$ 646.1 MILLION

1954

AMERICAN AIRLINES

American Airlines expanded its DC-7 service to Chicago and San Francisco in January when it received more DC-7 transports.

On January 10, the airline added a daytime westbound flight from New York to Los Angeles via Chicago, shaving an hour off the nonstop time between Chicago and Los Angeles in both directions. The eastbound trip from Los Angeles to Chicago took 5 hrs. and 30 mins, and the westbound journey from Chicago to Los Angeles took 6 hrs. and 10 mins. The new service also connected with a DC-6 in Chicago, providing a speedier service between Los Angeles and Washington.

AA also launched a DC-7 service from San Francisco to New York via Tulsa beginning on January 17. Due to the restrictions of American's certificate forbade it from flying nonstop services between San Francisco and New York (or Chicago), the Tulsa stop was made on both the eastbound and westbound flights.

American's third DC-7 service began on January 27 between Los Angeles and New York via Chicago. That flight was notable for providing an afternoon flight from Los Angeles to Chicago, departing the West Coast at 12.05 pm and arriving in Chicago at 7.30 pm.

DELTA-C&S AIR LINES

Delta-C&S's first DC-7 was christened at the Douglas factory in California on February 22 (the second arrived in March). The large propliner, carrying 50 passengers, piloted by Charles H. Dolson, vice president of operations, and T P Ball, superintendent of flight operations, flew from Santa Monica to Jacksonville on February 24 in 6 hrs. at an average speed of 371 mph (597 kph). The DC-7 was the first in a fleet of 10 that Delta-C&S had ordered at a total cost of $ 17.5 million. Six more were delivered before the end of 1954.

Consequently, after receiving its two DC-7s, the airline began scheduled operations on April 1 with two round-trip nonstop flights from Chicago to Miami (one deluxe and one coach). The flight time from ramp to ramp was 3 hrs. and 40 mins.

Delta-C&S Air Lines set a new commercial aircraft speed record on March 18 when one of its DC-7s flew from Santa Monica, California, to Jacksonville, Florida, a distance of 2,154 miles, in 5 hrs. 29 mins and 33 secs at an average speed of 391 mph (629 kph). The best previous time was on November 21, 1953, when a National Airlines DC-7 flew from Los Angeles to Miami, 2,341 miles (3,767 km), in 5 hrs. 50 mins, and 12 secs.

Delta-C&S Air Lines labeled their DC-7s the Golden Crown Fleet and emblazoned a crown, a symbol of power, on the enormous distinctive tail of each of its 69-passenger aircraft. The flights were named 'The Royal Biscayne,' 'The Royal Poinciana,' 'The Royal Ranger,' and the 'Royala Merchant'

NATIONAL AIRLINES

National Airlines received 3 of its Douglas DC-7s by January 5, while the fourth liner was in the yard at the Santa Monica factory and was delivered a couple of weeks later, fulfilling NAL's order.

The Miami-based airline launched its inaugural Douglas DC-7 flight to Washington on May 23. Then, on April 25, NAL launched a 3-hour, 30-minute nonstop service between Washington and Miami. The new DC-7 service shaved 20 minutes off the previous fastest time.

Three weeks later, on May 15, National started a DC-7 air coach service on its New York-Miami route, with a 3-hour 30-minute flight time. The DC-7 departed New York every night at 10 pm and was approximately 30 minutes faster than Eastern's Super Constellations. NAL operated the night flight to avoid interfering with the DC-7's First-Class daytime flights. Furthermore, the first-class 69-seating arrangement was retained for the coach operations.

In 1954, with the introduction of the 69-seat DC-7 coach liner services, NAL's spring-summer travel program provided more coach flights and seats between New York and Florida than any other 'off-season' time in the airline's history at 63.7%, which, coincidentally, was also the highest in the industry.

W A Patterson, president of United Airlines, celebrated his 25th anniversary with United Air Lines (20 years as president) on March 29, the same day the airline's first DC-7 was to be received from Douglas at Santa Monica. Thus, the intimate relationship between Douglas and UAL entered a new chapter.

On May 24, A brand-new UAL Douglas DC-7 flew from New York to Honolulu in 16 hrs. and 51 mins, including a single stop in San Francisco, breaking DC-7 records for each leg and the total distance. The overall flying time of the DC-7 from coast to coast was 8 hrs. and 17 mins. The special dawn-to-dusk flight to Honolulu carried media and airline officials. The airliner transported 37 passengers and 6 crew members from New York to San Francisco and picked up an additional 15 passengers and one crew member (making it 59 in total) before it departed for Hawaii.

United Air Lines began nonstop coast-to-coast service with its new Douglas DC-7s on June 1. The UAL DC-7 configuration included a cargo capacity of 9,000 lb. (3,149 kg), as well as the crew, 58 passengers, and baggage. The passenger cabin was divided into two sections, each with its own color scheme, and a special cabin-level forward compartment was created to store all passenger baggage in an upright position, thereby accelerating loading and unloading operations.

On June 1, United Air Lines launched a nonstop DC-7 service between San Francisco and New York, with a stopover in Denver. The daily 9 am (Pacific Daylight Time) flight departed San Francisco and arrived in New York at 7:35 pm (Eastern Daylight Time). The westbound flight left New York at 4 pm EDT and arrived in San Francisco at 10:30 pm PDT.

United Air Lines introduced DC-7s on all first-class flights to Honolulu on December 31. In addition, on that date, all of the remaining Stratocruisers were retired from service, and UAL's Boeing Stratocruiser fleet was sold to the British international carrier BOAC.

United Airlines placed an order for 17 DC-7s in September. With that purchase, the carrier's DC-7 fleet would expand to 27, bringing the total number of DC-7s sold by Douglas to 107. UAL also ordered 10 DC-6B Liftmasters and 5 DC-6A Liftmasters (freighters). It was a lucky month for Douglas.

4. United DC-7, N6302C, c/n 44266, 'Mainliner New York', flying above the clouds. DEL: April 20, 1954. NN: 'Mainliner Chicago'. LSD to US Overseas Airlines Jan 10, 1964. Sold to C & J Aircraft Leasing Jan. 24, 1964. RET 1969. WFU & STD at Ontario California, Sep. 1963. B/U 1968. (National Park Service)

5. Delta - C&S DC-7, N4872C, c/n 44262, number '702', DEL: March 25, 1954, at Chicago Midway Airport in 1954. Sold to BMR Aviation on June 30, 1966. WFU & STD at Ontario, California, Aug. 1966. B/U in 1968. (courtesy of Pat Bukiri and the Midway Airport Historians)

PAN AMERICAN WORLD AIRWAYS

In 1954, Pan American World Airways (PAWA) was the first airline to order the DC-7C. The $33 million order was for 15 aircraft for delivery in early 1956. PAA had also placed orders for 7 DC-7Bs.

The DC-7C was designed for a maximum gross takeoff weight of 140,000 lb. (63,503 kg) and 107,000 lb. (48,533 kg) on landing. Fuel capacity was 7,860 gallons (29,755 liters). Significantly greater than those of earlier DC-7 models, the weight increases were made possible by the increased climb power of Wright's imminent DA4 version of the Turbo Compound engine, as well as a substantial wing modification that made space for an increase in fuel capacity. The modification necessitated a 10-foot extension of the center section of the wing span. In addition, the wing modification reduced cabin noise to DC-6B levels by five feet by relocating the inboard engines further out. The wing modification necessitated substantial changes to the DC-7's control surfaces, such as a taller vertical stabilizer and a modified horizontal tail surface, to compensate for a more severe one-engine-out yaw problem.

'GREAT CIRCLE' ROUTE

PAWA expanded its year-old campaign for the 'Great Circle' route across the Pacific, leveraging the new long-range DC-7Cs, which could fly nonstop from Seattle to Tokyo. With the arrival of the Douglas DC-7Cs in 1956, PAA could fly the Pacific nonstop between Japan and the United States West Coast regularly using the short 'Great Circle' route.

In 1953, PAWA requested rights to the transpacific 'Great Circle' route, which, at the time, was controlled only by Northwest Orient Airlines. One faction in Northwest preferred selling or otherwise disposing of its 'Great Circle' rights, while another favored preserving the route and maximizing its traffic potential to a greater extent.

If allowed to fly across the 4,770-mile northern route to Tokyo, PAA's 15-plane DC-7C fleet would cut trans-Pacific travel time by more than half. PAA's route in 1954 was 6,688 miles (10,763 km) long, flying from San Francisco to Honolulu, Wake Island, and Tokyo. With the 'Great Circle' route's 5,222-mile (8,404 km) range and the DC-7C's cruising speed of 365 mph, PAWA could be the first to cross the Pacific nonstop both ways. With negotiations going on through 1954, PAA would have to wait for the 'Great Circle' route for a while longer.

DC-7B

6. An Eastern Air Lines DC-7B at Chicago Midway. N844D, c/n 45450. DEL: March 26, 1958. Sold to California Airmotive July, 27, 1965. LSD to Overseas National Airways (ONA) April 1966. RET in 1968. WFU & STD at Lancaster, California in 1968. BT by Allied Aircraft Dec. 11, 1970. Later B/U. (courtesy of Pat Bukiri and the Midway Airport Historians)

In 1954, flight tests were conducted on the DC-7B, the new intercontinental (long-range) version of the DC-7. The airliner first flew in May 1953.

The fundamental DC-7 configuration remained unchanged, but several changes and enhancements were made, increasing the DC-7's gross weight from 122,200 lb. (55,429 kg) to 125,000 lb. (58,514 kg) for the DC-7B. The improvements included an increased fuel capacity, a redesigned wing-flap linkage system for increased takeoff lift, and a 100-hp increase per engine from a later series of the Wright turbo compound engines.

The fuel capacity was increased from 5,512 gallons to 6,400 gallons (24,227 l) by adding saddle tanks to the upper wing surface as part of the larger engine nacelles. The new saddle-tank engine nacelles had a distinctly humped appearance, so the resulting drag had to be carefully monitored, particularly during takeoff with the open cowl flaps. In addition, the placement of these saddle tanks immediately behind the open exhaust stacks of the auxiliary turbines was surprising due to the possibility of a fire, even though the engine nacelles were primarily made of titanium alloy.

Due to the ability to carry an additional 90 gallons (340 l) of fuel in the center wing trailing-edge tanks, the takeoff gross weight increased to 125,000 lb. (56,699 kg), and the fuel capacity increased to 978 gallons (3,702 l). Douglas asserted that the DC-7B could fly nonstop from London to New York at a cruising speed of 360 mph (579 kph), making it the world's fastest piston-powered transoceanic propliner.

The ability of the DC-7's undercarriage to be lowered at airspeeds of up to 300 mph (483 kph) and utilized as an airbrake was an intriguing feature of the DC-7B. That was particularly useful while the airplane was in a 'holding pattern.'

In 1954, the DC-7B was powered by the Wright DAL turbo-compound engine, which produced 100 more METO horsepower than the previous DA2 engines (METO is the maximum power allowed to be produced continuously by an engine). This increase in horsepower increased gross takeoff weight to 125,000 lb. (56,699 kg). The first DC-7Bs were delivered to Juan Trippe's Pan American World Airways.

DC-7B Extended Range Specifications

POWERPLANTS	4 x WRIGHT 972TC18DA4 Turbo Compound Engines Takeoff: 3,250 hp / 2,900 rpm Normal : 2,700 hp / 2,600 rpm
CARGO CAPACITY lb. (kg)	17,700 (8,029)
MAX. SPEED mph (kph)	412 (663)
BEST CRUISE SPEED mph (kph)	360 (579)
CLIMB fpm (m/min)	525 (160)
SERVICE CEILING FT. (m)	27,900 (8,504)
TAKEOFF LENGTH ft. (m)	6,350 (1,935)
LANDING LENGTH ft (m)	5,870 (1,789)
MAX. STILL AIR RANGE mi. (km)	5,120 (8,240)
MAX. PAYLOAD RANGE mi. (km)	3,280 (5,279)
MAX. FUEL RANGE mi. (km)	4,920 (7,920)
EMPTY WEIGHT lb. (kg)	68,073 (30,877)
MTOW lb. (kg)	126,000 (57,153)
MAX. LANDING WEIGHT lb. (kg)	102,000 (46,266)
WINGSPAN ft (m)	117.6 (35.84)
LENGTH ft. (m)	108.11 (32.95)
HEIGHT ft. (m)	28.7 (8.74)
SEATING	69-99

DC-7 OPERATORS 1ST QUARTER 1954 (USA DOMESTIC)

AIRLINES	DC-6/ A/B
AMERICAN	7
NATIONAL	8
TOTAL	**15**

DC-7 OPERATORS 1ST QUARTER 1954 (USA INTERNATIONAL)

AIRLINES	DC-7/ 7B
AMERICAN	12
NATIONAL	8
TOTAL	**20**

7. On a wet and gloomy day, a United DC-7 at Chicago Midway Airport is taxiing to its takeoff position and is about to pass a Trans-Canada Air Lines 'Canadair North Star' and a Braniff Convair Liner. (The North Star was also known as the DC-4M and, in the UK, as the Argonaut. It was powered by Rolls-Royce Merlin V12 engines, giving it a speed of 325 mph / 523 kph) compared to the standard DC-4's cruising speed of 227 mph / 365 kph). (courtesy of Pat Bukiri and the Midway Airport Historians)

8. Continental Airlines DC-7B, N8211H, c/n 45194, 'City of Los Angeles'. DEL: April 5, 1957. Sold to California Automotive March 1, 1964. In May 1964 BT by AREA Ecuador, REG: HC-AIP named 'Guayas'. WFU & STD in Quito in 1968. Accident: In March 1966, the aircraft was destroyed beyond repair after it taxied into a ditch in Miami, Florida, no casualties. (courtesy of Pat Bukiri and the Midway Airport Historians)

INTERNATIONAL AIRLINES

SOUTH AFRICAN AIRWAYS

On August 6, South African Airways was the first international carrier to order the DC-7. The airline placed an order for 3 DC-7Bs. The aircraft were delivered early in 1956 and were slated for service on the Johannesburg-London route. The order, including spares, was worth $8,400,000. The South African order raised the number of DC-7s ordered or sold by the Douglas company to 108.

DC-7C

In late June, Douglas Aircraft Company proposed to the airlines a new version of the DC-7 with a range of 5,000 miles (8,047 km). The DC-7C was a new kind of aircraft that differed from previous Douglas four-engine commercial transports primarily in having a 10-foot (3 m) larger span, resulting in a wing area of 1,642 sq. ft. (152.54 sq. m.). It was equipped with saddle tanks with the Wright Turbo Compound DA4 engine and was fitted with Hamilton Standard propellers. The DC-7C's gross takeoff weight was 137,960 lb.(62,578 kg), whereas the DC-7B's was 126,000 lb. (57,153 kg). The airframe was extended and weighed 140,000 lb. (63,503 kg) gross, while its maximum landing weight was 107,000 lb. (48,534 kg) compared to the DC-7B's 102,000 lb. (46,266 kg). Deliveries of the DC-7C started in the fall of 1956.

THE LAST OF THE PROPLINERS

The Douglas Aircraft Company designed and built its first multi-engine airliner in 1932. It was designated the DC-1 (DC stood for Douglas Commercial). Subsequently, thousands of airliners were created and constructed by the Douglas Company between 1932 and 1954. After the DC-1 came the DC-2, then the immortal DC-3, which still takes to the skies, the sturdy C-54/DC-4, which served admirably in WWII, the smooth operator, the DC-6B, the best-selling Douglas propliner of all, and lastly the King of the Seas, the DC-7 'Seven Seas.'

Unfortunately, after 22 years of producing exceptional piston-engine airliners, Douglas announced that the DC-7 would be the company's final piston-powered airliner. The aircraft manufacturer intended to concentrate on the Douglas DC-8 jetliner. The jet age effectively ended one of the greatest Douglas propliners, the Douglas DC-7C.

TECHNICAL DESCRIPTION OF THE DC-7C 'SEVEN SEAS'

The four-engine Douglas DC-7C exemplified excellence in enhancing something that had already been proven effective. Since 1945, with the DC-4, then with the DC-6 and DC-7, Douglas had met the global demand for dependable, robust, and elegant long-range propeller aircraft. However, there was a problem; none of these three aircraft types had sufficient range to provide year-round, nonstop transatlantic flights in both directions.

In June 1954, Douglas began investigating methods to improve the DC-7's basic design with sufficient range for transatlantic service. The result was the DC-7C. In 1956, Pan American began transatlantic service with the DC-7C, nicknamed the 'Seven Seas.'

The first DC-7 was distinguished from the DC-6 by its extended fuselage and the Wright Turbo-Compound engines that first appeared in 1953. The DC-7 propliner was designed predominantly for the US domestic market, and a version for overseas operations was already in development before the DC-7 entered service.

The overseas version, the DC-7B, featured several design refinements that accompanied the increased fuel capacity and total weight. The modifications included an extension aft of the nacelles (to accommodate additional fuel storage), which decreased their drag coefficient, and a redesigned wing-flap linkage for improved efficiency to keep the takeoff distance down at the new gross weight. The DA-4 variants of the Turbo-Compound engines provided a modest increase in power.

The dilemma for the manufacturer was how to increase the fuel capacity without incurring severe penalties. Douglas had the option of adding wingtip tanks and underwing tanks. However, the company chose a second option: extending the wing root to increase the wingspan.

The wing root solution, which was not straightforward from a production standpoint, had several advantages. A 5-ft. (1.52 m) section in each wing between the inner nacelle and the fuselage gave ample space for the extra 1,124 gallons (4,255 l) of fuel required to attain the specified range. At the same time, additional wing area was made available to compensate for the inevitable increase in gross weight. Furthermore, the engines were relocated further away from the fuselage, resulting in a decrease in cabin noise.

The modifications increased the DC-7's weight from 126,000 lb. (57,153 kg) on the DC-7B to 143,000 lb. (64,864 kg) on the DC-7C. Therefore, more powerful variants of the Turbo-Compound, the 'EA-1', were made available.

Douglas lengthened the fuselage by an additional 3 ft. 6 inches (1.1 m) to accommodate the new weight, necessitating structural modification. Douglas also increased the height of the tail fin and rudder by some 2 ft. (0.6 m) to provide an increased area to counteract the added yawing moment in the asymmetric thrust condition with one engine inoperative. In addition, because of the increased power, larger ailerons and flaps were utilized.

The DC-7 was famous for being the first commercial aircraft to use the Freon ground and flight auxiliary cooling system and the first to incorporate titanium on a considerable scale in the power plant. These features were carried over to the DC-7C.

Also utilized in the semi-monocoque fuselage structure was titanium. The wing was a three-spar design with integral fuel tanks outboard of the undercarriage; there were 22 saddlebags on each side, including one on top of each nacelle.

The fuselage was pressurized to a differential of 5.46 psi using engine-powered superchargers in the outer nacelles.

POWERPLANT

Four Wright Turbo-Compound engines provided propulsion for the 'Seven Seas.' The new EA-1 version of the proven Wright Turbo Compound engine featured a takeoff rating of 3,400 hp and a guaranteed maximum cruise power of 1,910 hp at 11,600 ft. (3,536 m) in low ratio and 1,800 hp at 20,600 ft. (6,279 m) in high ratio. Although there were significant design modifications, the engine was designated as Model '988TC18EA-1', roughly corresponding to the R3350-32 military engine. The engine was in an advanced design stage, with the first units being constructed in late 1954.

The EA-1 was a logical progression of the Turbo Compound engine series DA-3 and DA-4. Changes were made to the design to enhance power and durability.

THE DC-7's DEVELOPMENT

On December 20, 1955, the first Douglas DC-7C made its maiden flight in Santa Monica. The aircraft was in the air for 2 hrs. 45 mins. It was soon followed into the sky by additional DC-7Cs coming off the assembly line.

Douglas established a certification deadline of July 1, 1956, and 3 aircraft were allocated to the test program, which was completed without a hitch in record time, six weeks ahead of schedule. On May 17, 1956, Douglas announced receipt of the CAA airworthiness certificate, by which time 4 production aircraft had already been delivered to Pan American Airways. At the same time, Douglas reported an all-around performance improvement over the original specification.

SEATING Every airline had a unique layout. However, the standard Douglas configuration accommodated 62 passengers in three compartments with four-abreast seating.
Five-abreast seating for up to 80 passengers with a 38-inch pitch (0.96 m) was offered for tourist class. The seats were on tracks for optimum flexibility, and movable bulkheads were provided.

START OF PRODUCTION

The DC-7C 'Seven Seas' entered production in September. The DC-7C had a range of almost 5,000 miles and could fly westbound from Europe to New York with a full payload, even in high winds. This allowed transatlantic carriers to reduce travel times from 3 to 5 hours on westbound flights and from 1.5 to 3 hours on eastbound flights. The cruise speed of the DC-7C was 350 mph (563 kph). America's world airline, Pan American World Airways, ordered 15 DC-7Cs, with the first 6 arriving in June 1956.

INTERNATIONAL AIRLINES

SAS

In late 1954, Scandinavian Airlines System (SAS) was the second non-US airline to order the DC-7C. The carrier paid Douglas $24 million for 8 Douglas DC-7Cs for delivery in the summer of 1956. The deal boosted the Scandinavian company's total financial spending for aircraft engines and spare parts in the United States to more than $114 million since 1946.

The SAS order and a breakdown of expenditures demonstrated the importance of international airline companies to US aircraft manufacturers. SAS, for example, had 14 DC-6Bs, 12 DC-6s, 4 DC4s, and 10 DC-3s in 1954, all Douglas airliners that formed the backbone of SAS' fleet. While the initial SAS orders were primarily Douglas/Pratt & Whitney business, apart from the many part suppliers, the DC-7C order in 1954 included Curtiss-Wright's R3350 Turbo-Compound engines, introducing the US engine manufacturer to the international markets.

The DC-7C enabled SAS to shave at least two and a half hours off the current 14-hour and 35-minute average New York-Copenhagen flight, completing the 3,846-mile (6,190 km) journey in 12 hours flat. The DC-7C also boosted SAS's seating capacity on first-class Royal Viking flights from 34 to 38 and the airline's tourist 'Globetrotter' service from 71 to 76.

SWISSAIR

The first European carrier to order the DC-7C was Swissair. The airline ordered 2 for delivery in late 1956. Swissair, like SAS and KLM, was a firm Douglas devotee. One would think if Boeing offered them 377 Stratocruisers for free, they would look the other way. That was how strong the connections those three carriers had with the Douglas Company were. In fact, KLM operated every Douglas/McDonnell-Douglas model the company produced, from the DC-2 in 1934 right up to the DC-10-30 in 1972 (including the DC-5, in which KLM was the initial customer. Only 5 DC-5s were manufactured).

9. An SAS DC-7C on short final to RWY 31 to Bromma Stockholm Airport, Sweden in 1965. REG: OY-KNC, c/n 44932, 'Odd Viking', DEL: Oct. 4, 1956. BT by Sud Aviation (France) on Aug. 11, 1967. WFU & STD at Bordeaux, France. BT by Air Management Corp. Aug. 13, 1974. B/U at Bordeaux in 1980. (Lars Söderström)

10. Swissair DC-7C, HB-IBP, c/n 45553, 'Schwyz' DEL: Nov. 4, 1958. At Zurich-Kloten Airport, Switzerland, in 1968. This was the last DC-7C to leave the Douglas plant. WFU & STD March 1962. Later it flew with SAS, Japan Air Lines, and ended up with the Spanish leisure airline, Spantax from 1965 to Jan.1977. WFU & STD. Later, in 1979, it was bought and turned into a restaurant. (ETH Zurich)

DC-7C SEVEN SEAS SPECIFICATIONS

POWERPLANTS	4 x WRIGHT 98R8TC18EA4 Turbo Compound Engines Takeoff: 3,400 hp / 2,900 rpm Normal : 2,800 hp / 2,600 rpm
CARGO CAPACITY lb. (kg)	18,440 (8,364)
MAX. SPEED mph (kph)	406 (653)
BEST CRUISE SPEED mph (kph)	371 (597)
CLIMB fpm (m/min)	507 (155)
SERVICE CEILING FT. (m)	21,700 (6,615)
TAKEOFF LENGTH ft. (m)	5,590 (1,704)
LANDING LENGTH ft (m)	5,360 (1,634)
MAX. STILL AIR RANGE mi. (km)	6,010 (9,672)
MAX. PAYLOAD RANGE mi. (km)	4,605 (7,410)
MAX. FUEL RANGE mi. (km)	5,640 (9,075)
EMPTY WEIGHT lb. (kg)	72,643 (32,917)
MTOW lb. (kg)	143,000 (64,865)
MAX. LANDING WEIGHT lb. (kg)	109,000 (49,442)
WINGSPAN ft (m)	127.6 (38.89)
LENGTH ft. (m)	112.3 (34.22)
HEIGHT ft. (m)	31.10 (9.48)
SEATING	69-99

11. The Wright '98R8TC18EA4' Turbo Compound Engine weighed 3,675 lb. (1,667 kg). A total of 159 were built between Sep. 1957 to July 1958. This engine was specifically made for the DC-7B, and DC-7C. Here, two Swissair mechanics are seen working on the port side inboard engine. (ETH Zurich)

1955

DOUGLAS

In 1955, the Douglas company expanded its production and office facilities, augmented the number of models under development or production, and extended the size of its personnel during its 35th year of continuous operation under the same management team.

Despite receiving record orders in the commercial airliner division in 1955, Douglas' backlog in that segment was less than 25%. Over 75% of Douglas' business was for various US Air Force and US Navy aircraft and missiles for the US Army and Navy (as well as exports to NATO Air Force countries).

From the start of 1955, Douglas received many orders for commercial propliners. Seventeen international airlines purchased/ordered 109 four-engine transport aircraft in the first two months. Before the end of 1955, the company received orders from 30 airlines for over 230 propliners, which was an all-time high for the industry. They included the DC-6A cargo planes, DC-6Bs, DC-7s, DC-7Bs, and DC-7Cs. Out of that number, more than 100 orders were for the DC-7C.

THE DC-7C v THE L-1649A

The introduction of the Douglas DC-7C 'Seven Seas' in 1954 caused a stir in the aviation industry; the airliner made flying nonstop across all oceans and continents possible. Nonetheless, the Douglas Corporation had one competitor. The Lockheed Aircraft Company released the L-1049G on December 17, 1954. Its range was substantially more significant than that of the L-1049E, which was widely used at the time. It was more luxurious in its layout but slower and less economical to operate than the DC 7C. The L-1049G could not compete on international routes with the 'Seven Seas.' Lockheed had a problem; It had to make the L-1049G more economical and faster than it was or create a new airliner altogether.

Before Lockheed designed the L-1049G, it had worked on developing a Super-Constellation powered by the Pratt & Whitney T34 engine. The aircraft was designated R7V-2, and the airlines were naturally interested. However, the R7V-2 was a military development. The military went cold on the project and canceled it because the T34 powerplant was too complex and could only be used on larger transport aircraft such as the Douglas C-133 Cargomaster. Only 4 R7V-2s were built before the project turned to dust.

Lockheed understood it had to provide airlines with an airplane that could genuinely compete with the 'Seven Seas.' The situation worsened when Pan American Airways chose the DC-7C for the critical transatlantic route. PAA was quickly followed by Swissair, SAS, Sabena, and BOAC, all of which ordered the DC-7C.

All eyes were on TWA, a long-time operator of Lockheed aircraft. Every day came additional contradictory information from the company, such as Howard Hughes had selected the DC 7C. At the same time, other sources said that he had directly participated in Boeing 707 tests, and even more, Hughes was interested in the L-1549 turboprop Constellation (which was a non-starter). Hughes was causing headaches for Lockheed with his indecisiveness. Only one thing was certain: the trailblazing airline run by the shrewd Juan Trippe, PAA would put the DC-7C on the Atlantic in June 1956, and TWA would also have to put something equivalent online, or it would be behind the eight ball. The situation was no different for all other airline companies operating on the Atlantic run.

On April 19, the news that everyone had been waiting for had arrived. Howard Hughes ordered 24 L-1649 Super-Constellations, a new type of propliner with a speed equivalent to and a range more significant than any other piston engine transport manufactured at the time.

From then on, airline operators who had not yet decided on fleet renewal had an option between two aircraft of relatively comparable class: the DC 7C or the L-1649 Super-Connie. How did these two planes stack up?

The cost of a DC-7C was $2,295,500, while the L-1649 went for $2.4 million. Both airliners were powered by 4 Wright R-3350 turbo-compound engines rated at 3,400 HP each. Regarding the propellers, the DC-7C had four-bladed props, 13.94 ft. (4 m 25 in) in diameter, and the L-1649's propellers were larger at 16.83 ft (5 m 13 in).

The Lockheed L-1649's revised wing allowed the engines to be spaced out more than on earlier models. The plane's propeller diameter was raised higher, and the reduction ratio was reduced accordingly.

Regarding performance, the L 1649 had a maximum takeoff weight of 155864 lb. (70,700 kg), whereas the DC 7C had a maximum takeoff weight of 138,89 lb. (63,000 kg). The distinction was noticeable. This was owing to the L-1649's 'high performance' wing, which had a much higher structural weight. Depending on its fuel reserves and at the same load, the L-1649 would naturally go further than the DC-7C. That would be the case on transatlantic routes, which remained the most competitive.

Regarding the DC-7C, it did not have the range of the L-1649, but it was cheaper and faster than the L-1649. Ultimately, these figures finally mattered to Douglas and Lockheed: DC-7Cs produced: 338, L-1649s produced: 44. Douglas won the battle of the last two piston engine giants.

UNITED AIR LINES

The final DC-7 in United Airlines' initial order of 25 was delivered in late January at Douglas' Santa Monica facility. Plane No. 1 was delivered a year earlier, in March 1954. In 1956, United received two more DC-7s for a $22,300,000 deal, including 5 DC-6A all-cargo planes and 10 58-passenger DC-6Bs.

On May 6, United Air Lines became the first airline to operate nonstop flights in both directions between New York and San Francisco when it launched a New York-San Francisco DC-7 nonstop service, shaving an hour off the company's current one-stop service between the two cities. The revised schedule provided UAL with two daily nonstop flights between New York, San Francisco, and New York and Los Angeles.

Also in May, UAL placed a large order for 26 four-engine Douglas aircraft, totaling around $42.5 million. These planes, together with another 17 already on order, provided United with 43 additional four-engine airliners for a total cost of $64,800,000 to support projected expansion into the late 1950s—the order called for 15 DC-7s at $1,902,000 per plane and 11 DC-6Bs.

12. United Air Lines DC-7, N6312C, c/n 44276, 'Mainliner Newark' DEL: Aug. 4 , 1954. This DC-7 was from the initial batch of 25. Sold back to Douglas, on Feb. 28, 1960. BT by California Automotive in Sep. 1963. Sold to TASSA (Spain) in April 1964 REG: EC-WAE. RR July 1965 REG: EC-BAE. WFU & STD at Palma, Majorca, Spain in 1965. B/U. (courtesy of Pat Bukiri and the Midway Airport Historians)

EASTERN AIR LINES

Eastern Air Lines placed an order for eight more Douglas DC-7Bs in February. The company planned to expand coach services to 70% of its total operations. The eight new planes were scheduled to arrive in 1956. These planes and another 17 already on order provided EAL with 43 additional four-engine airliners for a total cost of around $64,800,000 to support projected expansion into the late 1950s.

As the new DC-7Bs were incorporated into EAL's full-fare express service, E. V. Rickenbacker, EAL's colorful character, World War I ace aviator, Medal of Honor recipient, and board chairman, relegated the Super Constellations to fly on expanded coach operations. Consequently, by the end of 1956, EAL had 48 super-air coaches in its fleet. In total, the carrier had 119 aircraft at its disposal that year.

'THE GOLDEN FALCON'

EAL, which had long promoted its 'Great Silver Fleet,' added a 'golden' theme to its new Douglas DC-7Bs. The slogan, 'The Golden Falcon DC-7B Built by Douglas', greeted boarding passengers at the doorway.

The DC-7B's cabin interiors were richly trimmed in gold to emphasize the' golden' theme. The first 'Golden Falcon' was delivered to EAL in early June, and two more followed at the end of that month.

RE-EQUIPMENT PROGRAM

Eastern ordered 20 more DC-7Bs in early August. Eastern's DC-7B 'Golden Falcon' fleet increased to 40 with the additional order, in which the DC-7Bs were delivered between October 1957 and April 1958. Eastern had received 5 of the original 20 DC-7Bs ordered, but by August 1956, they had all been delivered to the carrier. By August 1956, they had all been delivered to the carrier.

Eastern's re-equipment program predated the arrival of the DC-7s. The program began in 1950 and featured ground installations in addition to airliners. The overall cost of the program was an astounding $350 million.

13. An EAL 'Golden Falcon' DC-7B at Chicago Midway in Sep. 1959, N845D, c/n 45451, DEL: March 31, 1958. Sold to California Automotive Dec 7, 1964. Then to Transair Sweden AB on June 13, 1965 REG: SE-ERK. WFU & STD at Malmo Sweden in Nov. 1968. The following year it was B/U. (courtesy of Pat Bukiri and the Midway Airport Historians)

AMERICAN AIRLINES

American Airlines ordered 14 additional DC-7s for delivery in late March between 1956 and 1957. The new order, added to the 25 DC-7s already delivered to American, gave America's largest airline a total of 39 of the turbo compound-powered propliners. American Airlines had a total fleet of 191 aircraft at the time of the DC-7 order: 25 DC-7s, 50 DC-6s, 25 DC-6Bs, 7 DC-6As, 75 Convairs, and 9 DC-4 freighters.

14. American Airlines DC-7B, N348AA, c/n 45401, DEL: Sep. 20, 1957, at Chicago Midway in 1959. Destroyed: This DC-7B was on a supply flight during the war in Biafra, Nigeria in 1969. It was flying under the Swedish Red Cross which purchased the aircraft in March 1969. The DC-7B was shot down by a Nigerian Air Force MiG-17 jet fighter on June 5, 1969. All 4 Swedish occupants in the aircraft perished in the ensuing crash. (courtesy of Pat Bukiri and the Midway Airport Historians)

Braniff International Airways (BA) ordered 7 DC-7Cs worth $20 million. The order was the first for new long-haul equipment placed by Braniff in several years. It was also the airline's most significant investment ever made for flight equipment, which included acquiring spare parts and special shop equipment for maintenance. Braniff was scheduled to receive the initial 3 turbo compound-powered propeller aircraft in the autumn of 1956, with the remainder arriving in the first half of 1957.

Braniff's DC-7Cs were configured to accommodate either 62 first-class passengers or a combination of 75 first-class and tourist passengers. With the latter configuration, the forward tourist section accommodated 59 passengers, while the rear compartment accommodated 16 first-class passengers. Some aircraft were also deployed in an all-tourist configuration with 81 seats.

The DC-7Cs were to be initially utilized by Braniff on its South American routes, which were serviced by DC-6 aircraft. In 1955, Braniff's fleet consisted of 9 DC-6s, 24 Convair 340s, and 24 DC-3s. The introduction of DC-7Cs strengthened the airline's competitive position in South America, where it confronted stiff competition from Panagra's DC-7Bs. Moreover, Braniff's other competitor in the South American market was Pan American World Airways, whose fleet consisted of DC-7Bs and DC-7Cs.

NORTHWEST AIRLINES

15. Braniff DC-7C at Chicago Midway. N5906, c/n 45074, DEL: May 16, 1967. BT by Charlotte Aircraft Corp. on Sep. 15, 1970. WFU & STD at Ryan Field , Arizona in Sep. 1970. Later B/U. (courtesy of Pat Bukiri and the Midway Airport Historians)

Northwest Airlines (NWA) raised its order for new Douglas aircraft from 10 to 14, with 8 DC-7Cs and 6 DC-6Bs. The propliners were scheduled to be delivered between January and October of 1957. The total order was $27 million. NWA announced its order for 10 new aircraft from Douglas in March, but a decision regarding the aircraft type was postponed until April 30, pending the results of a recent management survey. This order was NWA's first for DC-7s.

16. A Northwest DC-7C makes its way to the gates at Chicago Midway in September 1960. Note the TWA Connie on the left and the American L-188 Electra, and North Central Airlines Convairliner on the right. (courtesy of Pat Bukiri and the Midway Airport Historians)

PAN AMERICAN WORLD AIRWAYS

PAWA, which had previously placed an order for 7 DC-7Bs in 1953, expanded its DC-7 orders from 7 to 40 in April 1955, bringing its total investment in new piston-engine aircraft to $110 million. The new order included 18 DC-7s, in addition to 15 DC-7Cs ordered in July 1954 and the 7 DC-7Bs mentioned above. The first DC-7B was delivered to PAA at the end of May 1955. In addition, the first DC-7Cs were delivered in early 1956. Both propliners were powered by the Wright Turbo Compound engines. Pan Am's DC-7Bs were used for nonstop air-coach flights between New York, Paris, and London and similar nonstop services on Pacific routes.

TRANSATLANTIC TRAVEL TAKES OFF FOR PAWA

The completion of Pan American World Airways' 50,000th transatlantic flight on June 28, 1955, highlighted two significant facts in this field. Since Pan American pioneered the first scheduled passenger service across the Atlantic in Boeing flying boats in 1939, fantastic progress was made in connecting Europe and North America by air. Second was the mid-1950s air travel boom across the Atlantic, breaking numerous coach and first-class travel records. It must have seemed long ago when, in 1939, the twice-weekly flying boat service from Newfoundland to Europe typically took 24 hours. Just 16 years later, Pan American dispatched 116 weekly flights to Europe using Boeing 377 Stratocruisers, Douglas DC-6s, and DC-7Bs that could fly nonstop to London in 12 to 14 hours. In the final week of June 1955, statistics showed that 569 US and international airlines arrived from or departed for Europe at New York's Idlewild Airport.

For the second consecutive year, more passengers traveled to Europe by air than by steamships. Moreover, the record of 580,000 trans-Atlantic airline passengers set in 1954 was surpassed in 1955 by over 600,000 passengers, primarily due to the efforts of Pan American World Airways.

PAWA's JUNE RECORD

Pan American World Airways created a new record when it carried 14,760 passengers between the US and Europe throughout June 1955. Demand for seats on Pan Am's new DC-7Bs contributed to the increase in traffic volume. The Peak was reached during the last week of June when approximately 3,900 passengers were flown from the US to Europe, compared with 2,913 persons traveling on eastbound PAWA flights during the same period in 1954. The carrier's transatlantic propliners logged more than 50 million passenger- miles during the record month.

17. PAWA DC-7B, N776PA, c/n 44870, 'Clipper Nonpareil' DEL : Aug. 17, 1955. BT by Cross Aero Corp. Oct. 1964. Sold to California Airmotive in April 1965. BT by Aerolineas Peruanas Aug. 1965, REG: OB-R-784, WFU & STD at Miami, Florida July 1968. Later B/U. This Pan Am DC-7B has the iconic 'Globe' livery which was introduced in 1958. This scheme was designed by Edward Larrabee Barnes and Charles Forberg. (The University of Miami Libraries)

PAWA WINTER FLIGHTS

In September, all seven of Pan Am's DC-7Bs were placed into service. This action was taken in preparation for the winter schedule that began on October 30. Pan Am increased its DC-7B transatlantic flights during the winter season to 14 round trips per week. Three weekly flights were made available, with first class and tourist class accommodation. The route was as follows: New York-Shannon-Paris-Rome-Beirut-Damascus-Tehran. Additionally, all direct tourist flights to London, Paris, and Rome were operated by DC-7Bs. Other cities provided with Pan Am DC-7B tourist flights in the winter of 1955 were Amsterdam, Frankfurt, Dusseldorf, and Istanbul.

DELTA-C&S AIR LINES

Delta-C&S Air Lines ordered five more Douglas DC-7s on May 12, with the first slated to arrive in December, in time for the carrier's winter services.

Delta-C&S ordered six additional DC-7s in August, with deliveries commencing in 1957. The latest order brought the total number of DC-7s Delta had on order to 11. One of the 11 ordered DC-7s was delivered in the fall of 1955. After the DC-7 deliveries were completed in 1957, there were 21 DC-7s in Delta's fleet.

PANAGRA

PAWA affiliate Pan American Grace Airways began twice-weekly Douglas DC-7B service between Miami and Buenos Aires on August 1, cutting 1.5 hours from current DC-6 schedules. Panagra's DC-7Bs could carry 58 passengers and had a lounge, bar, and 15 bed-sized berths. Moreover, the DC-7s were equipped with weather radar. The flight to Buenos Aires took 16.5 hours.

PANAGRA DC-7 RECORD

Panagra's DC-7B pre-inaugural El Interamericano flight to South America (service started on August 1) set a new airline record of 20 hrs 2 mins elapsed time (actual time an airliner spends in the air) on

18. Two Delta DC-7s at Chicago Midway Airport in July 1961. The front plane is N4873C, c/n 44263, no. '703'. DEL: March 25, 1954. BT by BMR Aviation Aug. 3, 1966. WFU & STD at Ontario, California June 1966. B/U in 1968. (courtesy of Pat Bukiri and the Midway Airport Historians)

the return flight from Buenos Aires to New York. The average speed was 329 mph (529 kph), and elapsed time was approximately 3 hours below that for regular service between the two points.

The pre-launch flight of Panagra to South America was a mission of goodwill. It carried 53 passengers, including senior Civil Aeronautics Board (CAB) officials, 12 media personnel, 15 flight attendants from eight different airlines, a group of Panagra executives, and technical staff from Douglas Aircraft and Wright Aeronautical. The eight-day program included meetings with government leaders in Panama, Peru, Chile, and Argentina, as well as news conferences, wreath-laying ceremonies, luncheons, receptions, and other similar activities.

DOUGLAS EVENTS

DOUGLAS REARRANGES ASSEMBLY LINE

In February, Douglas-Santa Monica began restructuring its manufacturing line to accommodate the DC-7C while the manufacture of DC-6/7 models proceeded.

The aircraft manufacturer invested $1 million on the conversion and expansion program, which included the construction of 5 new subsidiary buildings for the long-range DC-7 engine uncrating, engine buildup, maintenance storage, electrical supply, shops, and other facilities.

At Santa Monica, tooling was put together to allow the DC-6A, DC-6B, DC-7, DC-7B, DC-7C, and the USAF's C-118A to all come down the same final assembly line.

DOUGLAS BACKLOG

Douglas had a commercial order backlog of 274 DC-6/7 series aircraft in August. The aircraft on order were: 17 DC-6As, 85 DC-6Bs, 31 DC-7s, 42 DC-7Bs, and 99 DC-7Cs.

Up to that month, the company had already delivered 462 DC-6/7 series aircraft, which included 174 DC-6s, 25 DC-6As, 186 DC-6Bs, 64 DC-7s, and 13 DC-7Bs.

INTERNATIONAL AIRLINES

BOAC

THE DC-7C v BRISTOL BRITANNIA SAGA

The British Ministry of Transport reported on Monday, January 10, 1955, that the Transport Minister had received a letter from BOAC approving the purchase of 19 Douglas DC-7Cs. However, the airline neither confirmed nor denied the claim. Yet, when government officials learned the news, some were not happy.

The decision as to whether BOAC required 19 DC-7Cs to maintain its competitive position was primarily based on a performance and availability comparison between the Douglas DC-7C and the British 4-engine turboprop, the 'Bristol Britannia.'

The DC-7C's key selling point was its capacity to perform nonstop transatlantic services in both directions with high frequency while carrying an economic payload. As a result, the British believed that an airplane with that speed could claim the majority of first-class transatlantic traffic, relegating the existing airliners to tourist services with at least one stop. By the end of 1955, BOAC's North Atlantic flights were only operated by Stratocruisers, which could not fly westbound nonstop but could occasionally fly eastward nonstop with the help of strong tailwinds.

BOAC's ability to compete with American DC-7Cs and long-range Super Constellations was totally dependent on the Britannias that the carrier had already placed on order. The first 15 Britannias manufactured for BOAC were the Mk 100 type (powered by Proteus 705 turboprop engines), which were not considered capable airliners (the Britannia Mk. 100, like the DC-7, was incapable of flying across the North Atlantic).

The remaining 10 BOAC Britannias were Mk 300s (powered by more powerful Proteus 755 turboprop engines) and were 'Long-Range' models with larger fuel tanks. Bristol claimed the Britannia 300 'LR' was superior to the DC-7C on all counts. The British turboprop had a 20% advantage over the DC-7C in both eastbound and westbound directions. The Britannia 300 'LR' became available in 1956.

The size of the DC-7C order proposed by BOAC came as a surprise. Many in Britain preferred that BOAC purchase solely British airliners, but they would understand if BOAC purchased the DC-7C. After all, Capital Airlines' huge Vickers Viscount purchase had been welcomed by the Americans. Many individuals in the UK said they would support BOAC's acquisition of American airliners if it were proven to be in the British flag carrier's best interests.

Many others, on the other hand, claimed that BOAC did not need as many as 19 DC-7Cs in order to guarantee competitive North Atlantic flights in 1957 and that regardless of the advantages of these propliners, they were unlikely to be competitive for passenger travel for more than a year or two. By then, the development, production, and early service of the Mk 300 'LR' turboprop Britannias would have gone smoothly, and they would be able to grab the bulk of the DC-7C's first-class traffic by 1958 while operating more cheaply than any piston-engine propliner.

On March 10, following consideration by the Board of BOAC of the British Government's authorization to order 10 (not 19) Douglas DC-7Cs, BOAC claimed that, despite recognizing the significance of this decision and the magnitude of the dollar expenditure permitted by the UK Government, BOAC's Board accepted it with appreciation and agreed. Thus, BOAC entered into new negotiations with Douglas for 10 DC-7Cs to be used until the British-made airliners could be brought into service economically on the transatlantic route.

SABENA

On Thursday, February 24, 1955, Sabena Belgian Airlines ordered six DC-7Cs. The first of the 62-passenger DC-7Cs was scheduled for delivery in the fourth quarter of 1956. Sabena used the aircraft to fly nonstop from Brussels to New York and from Brussels to Leopoldville in the Belgian Congo. This latest order brought the total number of DC-7 sales to 139.

Sabena ordered three additional DC-7Cs at a board meeting on September 1. The esteemed chairman of Sabena, Gilbert Perier, who marked his 25th anniversary with the airline on that day, made the decision to purchase the three DC-7Cs. With the earlier order for six DC-7Cs, Sabena's DC-7C fleet would increase to nine aircraft.

19. BOAC DC-7C, G-AOIC, c/n 45113. DEL: Nov. 27, 1956. Taking off from Manchester International on April 18, 1958 on BOAC's first scheduled non-stop daylight flight to NY (Idlewild). Sold to F B Ayer & Associates on Feb. 13, 1964, REG: N90801. BT by Saturn Airways on March 13, 1964. After going through several airlines, WFU & B/U in Miami, Florida in July 1985. (RuthAS)

KLM

KLM Royal Dutch Airlines ordered 10 Douglas DC-7Cs for delivery between April and July 1957. The order was valued at $31.5 million. These new aircraft joined KLM's four-engine fleet, which included 8 L-1049C Super Constellations, 10 L-749A Connies, 6 DC-6Bs, 2 DC-6As, 6 DC-6s, and 9 DC-4s. The Dutch carrier quickly followed up by ordering four L-1049Gs. This KLM purchase raised the total number of European airlines that had ordered the DC-7 series to five: BOAC, SABENA, SAS, Swissair, and KLM.

SWISSAIR

Another European carrier interested in the DC-7C was Swissair. The airline ordered two additional DC-7Cs on top of those placed in 1954. By 1955, the airline's fleet included 6 DC-6Bs, 7 Convair 240s, 3 DC-4s, and 9 DC-3s. With two DC-7Cs scheduled for delivery by the end of 1956, Swissair posed a significant challenge to its larger competitors for first- and tourist-class transatlantic traffic. Serving Europe and the Middle East, Swissair viewed its North Atlantic route as the most crucial in its network. Thus, to enhance its service on this vital transatlantic line, the Swiss airline opted to add aircraft with improved performance.

SAS

SAS, which had ordered 8 DC-7Cs in late 1954, increased its order to 14 in mid-1955. The new order totaled $42 million. In the spring of 1956, SAS received its first DC-7Cs, which were used on flights to New York and Los Angeles, along with other major intercontinental routes.

20. Swissair DC-7C, HB-IBL, c/n 45062 'Geneve', over Switzerland on March 31, 1959, DEL: Dec. 5, 1956. BT by Riddle Airlines Oct. 11, 1960, REG: N310G. It was converted to a DC7F in Dec. 1960. NN: Nov. 29, 1963, Airlift International. WFU & STD in Miami, Florida. B/U Nov. 1973. (ETH Zurich)

SAS' POLAR ROUTE

The general public received SAS's Polar Route between California and Europe well. The carrier's Polar flights to Europe were sold out for two months starting May 1955, the first tourist season for which Polar flights were made available. This route was served by a total of three flights per week in each direction.

The third Polar flight was introduced a month earlier than initially intended to accommodate the increase in passenger volume. That decision made 416 seats available per month, all of which were booked through July in the eastbound direction and September in the westbound direction. With SAS's launch of the Polar service in November 1954, nearly 1,600 passengers were transported in both directions.

SAS operated first-class and tourist-class flights with only one stopover after acquiring its DC-7Cs in 1956. However, in 1955, the SAS DC-6Bs that traveled the Polar route had two stops: Winnipeg, Canada, and Greenland. Clearly, the DC-7C was the superior option.

SAS planned to provide two types of tourist services in 1956: a mixed aircraft with 52 seats and 12 sleeper accommodations, and an all-tourist flight with space for approximately 75 passengers. As for first-class service, SAS's DC-7Cs would accommodate between 38 and 40 passengers. SAS also aimed to introduce a commercial route from Copenhagen to Tokyo via Anchorage, Alaska. These services were set to begin in early autumn with two to three flights per week.

SAS ORDERS MORE DOUGLAS AIRLINERS

SAS introduced a novel tourist configuration seating arrangement in 1955, replacing the previous straight line of three-abreast seating. The new design benefited SAS' DC-7C intercontinental tourist-class airliners. In this layout, the seats were staggered back on a slant so that the shoulder, armrests, and cushion junctions did not meet. Parallel rows were inclined across the space that was previously occupied by straight-line rows running to the rear of the passenger cabin.

21. SAS DC-7C, LN-MOE, c/n 44933, 'Reidar Viking', DEL: Nov. 13, 1956, seen here just after taking off from Stockholm-Arlanda Airport in May 1967. LSD by Nordseeflug March 21, 1968, REG: D-ABYF. NN in Aug. 1968 to Atlantis. RET to SAS March 21, 1969. WFU & STD at Kastrup-Copenhagen, Denmark, in April 1969. B/U May 1970. (Lars Söderström)

Called 'Comfortline' seating, the new configuration provided individualized seating without sacrificing passenger payload and cabin space. For the first time, passengers in the middle and aisle seats in the three-seat configuration had an unobstructed window view. SAS' DC-7Cs carried 44 tourist and 16 first-class passengers, with the latter being provided with sleepers. Cabins could be quickly re-configured to an all-tourist or all-first-class layout.

COMPANIA MEXICANA DE AVIACION

In 1954, Compañía Mexicana de Aviación (CMA), Mexico's largest airline and the world's third oldest airline, carried 450,000 passengers and flew 13 million miles. The airline became interested in offering nonstop flights between Mexico City and New York, leading to an order of three DC-7s for that purpose. The Mexican government expressed disappointment that hundreds of smaller countries had airlines operating into New York, while no Mexican carriers had scheduled flights there. As a result, it agreed to back the CMA project only if an agreement on the route could be struck on an exchange basis with the US government. This depended on the Civil Aeronautics Board (CAB) scheduling hearings regarding this route with American, Eastern, and Pan American. All three US airlines were also interested in flying nonstop from Mexico to New York. It was unlikely that the CAB would authorize more than one US carrier on the route, given that the traffic was around 50,000 passengers per year, which would hardly justify it, particularly since the load would need to be shared to some extent with the Mexican airline.

The most obvious of the three airlines from the Mexico side would be American Airlines, which already flew the route and had an extensive hangar facilities in Mexico, including Monterrey Airport. American Airlines had also invested much in increasing traffic to Mexico through advertising in the United States.

In the meantime, CMA, an affiliate of Pan American World Airways, ordered 4 DC-7C 60-passenger transports for delivery between 1957 and 1958. The new propliners reduced the flying time on the Mexican carrier's Los Angeles-Mexico City route by one hour.

DOUGLAS AIRLINER SALES 1955

MODEL	ORDERED	DELIVERED
DC-6	174	174
DC-6A	43	25
DC-6B	274	188
DC-7	116	64
DC-7B	56	23
DC-7C	105	0

1956

DOUGLAS

On August 6, 1956, Douglas delivered the first DC-7C to an international airline. SAS was the first European airline to operate the DC-7C. The Scandinavian airline named its DC-7C service the 'Global Express,' which was configured in the 'Royal Viking' luxury-service arrangement. SAS' DC-7Cs typically offered 40 passenger seats, or 60 and 79 for first-class and tourist operations, respectively. The DC-7C, the 'last of the big propeller-driven Douglas transports,' had a vast range that allowed airlines to conduct previously unattainable flights. By the end of 1956, 14 airlines had ordered 116 'Seven Seas' propliners.

The commercial model manufacturing line at Douglas' Santa Monica division was boosted throughout the year, and by the fourth quarter, it was producing 10 DC-6 and DC-7 series aircraft per month. By the spring of 1957, the rate had reached an all-time high of 15 per month.

Regarding American Airlines in 1956, the demand generated by the announcement of the introduction of 'The Royal Coachman,' a nonstop, coast-to-coast DC-7 aircoach flight, was so high before the inaugural flight on May 20 that AA had to schedule two round-trip flights per day, as opposed to the original one.

Nineteen fifty-six was also the initial year Braniff's extensive new fleet enhancement program was implemented. The first of Braniff's El Dorado DC-7C airliner, purchased under an $83 million new aircraft program authorized in 1955, was delivered to the airline in September and placed on schedules in late October. By December 1956, Braniff operated a fleet of 67 aircraft, including 4 Douglas DC-7Cs.

Late in 1955, Continental Airlines (CAL), like Braniff, announced a renewed fleet program to purchase 5 Douglas DC-7Bs, 15 Viscount 810D turboprops, and 4 Boeing 707 jetliners at a cost of over $63 million, including ground equipment and spares.

The airliner procurement program of Eastern Air Lines (EAL) included two types of piston-engine aircraft and two jet-powered types, including turboprops and pure jetliners.

Deliveries commencing in late summer 1956 and concluding in early December added 18 more luxurious 'Golden Falcons' to the Eastern fleet of Douglas DC-7B's and 10 Super G Constellations, equipped with radar and the newest fully automatic flight control systems.

These aircraft and the 12 DC-7Bs delivered to EAL in 1955 set up the airline to capitalize on the busiest fall and winter travel season between the northern United States and Florida in living memory.

The other airline in Florida, National Air Lines (NAL), was also undergoing a fleet expansion program with $ 108 million at its disposal; National placed orders with various aircraft manufacturers to deliver 6 Convair 440s, which were sent in the summer of 1956. EAL also ordered 4 Douglas DC-

7B's, 4 Lockheed Super H Constellations, 23 Lockheed L-188 Electra turboprops, and 6 Douglas DC-8 jetliners. The aircraft were all scheduled for delivery between 1959 and 1960.

Northwest Airlines, based in Minnesota, also ordered new piston-engine airliners in 1956. The company signed contracts for 26 new DC-6B and DC-7C aircraft and parts worth over $63 million. The aircraft deliveries began in late 1956 and continued through 1958.

DOUGLAS BACKLOG

In September 1956, the Douglas Aircraft Company delivered the 700th DC-6 and DC-7 series aircraft. The delivered airplanes had a total value of $750 million. The backlog for DC-6 and DC-7 airliners was 294 units, worth a hefty $500 million.

DC-7C GETS CERTIFICATED

On May 17, six weeks ahead of the original certification target date, the Douglas DC-7C received clearance from the CAA. This allowed the 'Seven Seas' to begin intercontinental services on June 1. Four aircraft had already been delivered to PAWA. The DC-7C took its first flight on December 20, 1955.

22. Cockpit of Douglas DC-7 of American Airlines 'Flagship Vermont', N334AA, c/n 45106. The aircraft was broken up in May 1965, but the nose section was sent to the Smithsonian Museum, central Washington DC, in Aug. 2010. (Smithsonian Museum)

23. Delta Air Lines DC-7B, N4890C, c/n 45354, no. '720' DEL: Jan. 6, 1958. BT by BMR Aviation March 31, 1968. WFU & STD Ontario California March 1968. BT by FM Enterprises May 1971. Sold to Ronald J Clark DEC. 1977. LSD by Columbia Pictures Feb. 1978. B/U Burbank, California April, 1978. Seen here at Chicago Midway in April 1959. (courtesy of Pat Bukiri and the Midway Airport Historians)

24. Delta DC-7 'Golden Crown', N4872C, c/n 44262, no. '702', DEL: March 18, 1954. BT by BMR Aviation on June 30, 1966. WFU & STD at Ontario, California, in June 1966. B/U 1968. Seen here at Chicago Midway on April 19, 1957. (courtesy of Pat Bukiri and the Midway Airport Historians).

Delta Air Lines (DAL) began operations on its new route between Washington and New York on February 1, with two of the three daily flights being low-fare coach options. The first was a DC-7 night coach service flying to New York, Washington, and Atlanta; the second service was on a DC-6, which flew to Fort Worth, Dallas, Atlanta, Washington, and New York; the final service was a DC-7 first-class flight serving New Orleans, Atlanta, Washington, and New York. DAL also served the New York metropolitan area from Newark Airport. The first-class flight was a crucial factor in the 1956 success of Delta Air Lines' new northeast routes. DAL's load factors were exceptional in its New York, Washington, and Atlanta services. In contrast, Delta's two daily coach flights had lower passenger numbers.

UNITED AIR LINES

On March 1, United Airlines launched an overnight nonstop DC-7 service from New York to Los Angeles. The service was named 'Starlight Continental.' The plane departed from New York at 12:30 am (EST) and arrived in Los Angeles around 6:15 am (PST). Passengers were offered a meal, followed by a nightcap, and could choose from sleep-inducing drinks such as warm milk or bouillon. The flight attendants (or stewardesses, as they were known back then) also provided passengers with sleep masks and slippers.

The DC-7 carried 8,000 lb. (3,629 kg) of mail, express shipments, freight, and luggage. UAL called the service the 'fastest overnight-to-Los Angeles cargo service in history.'

UAL ORDERS MORE DC-7s

UAL doubled its order for 15 Douglas DC-7s in April 1955, bringing its total DC-7 fleet to 57 aircraft. UAL was now the largest buyer of Douglas four-engine piston-engine transports, having ordered or received a total of 151 DC-6/DC-7 variants. United initially ordered 25 DC-7s, all of which were delivered by January 1955. A subsequent order for two additional DC-7s raised the total to 27, and a third order for 15 more DC-7s in 1955 increased the total to 42 aircraft. In April 1956, UAL received 27 DC-7s, leaving 30 still to be delivered.

THE SHORTEST DC-7 FLIGHT IN THE WORLD

The shortest DC-7 flight in the world in 1956 (and beyond) was United Air Lines' flight 734. The flight began in Washington, DC, and ended 25 minutes later in Baltimore. Furthermore, the following notice was included in the timetable for this particular flight: 'No local traffic may be carried on United flights between Baltimore and Washington,' with no explanation to the passengers. Eventually, the reason for the message was found to be that the DC-7 arrived in Washington the night before, laid over, and then flew to Baltimore the following morning and originated as 'Flight 745' flying to Washington, then on to Chicago and Seattle. The only traffic that could be carried was connecting business and any available mail and cargo. In one busy month in 1956, Flight 734 carried 40 passengers!

AMERICAN AIRLINES

American Airlines received its entire balance of 14 undelivered Douglas DC-7s in 1956. The first 5 DC-7Bs had the AA 'Royal Coachman' interior, accommodating 85 passengers plus a lounge. The DC-7Bs of American Airlines were equipped with Wright DA Turbo Compound engines, which were identical to the DA-2s except for minor instrumentation and electrical accessories. AA's DC-7Bs had a gross weight of 125,850 lb. (57,085 kg) and a landing weight of 102,000 lb. (46,266 kg).

DEFERRED AIRFREIGHT SERVICE

American Airlines' cargo division was also active in 1956. The airline offered a 'Deferred Airfreight Service' in May between Los Angeles, San Francisco, Chicago, New York, and Washington. This service was introduced on Sunday, May 6. Deferred Airfreight Service was provided on Douglas DC-7 Cargo aircraft and differed from conventional airfreight services in that shipments were typically not released from the destination location until four days after shipment.

The service was approved by CAB for one year and aimed to deliver faster than regular surface cargo at comparable rates.

ROYAL COACHMAN SERVICE

To meet the demand for seats, American Airlines launched its inaugural Douglas DC-7 'Royal Coachman' nonstop coach flight from Los Angeles to New York in two sections on May 20. Both sections were completely sold out. The additional section was introduced after the flight was fully booked two weeks prior to departure.

As the summer vacation season began in June, some scheduled flights were fully booked, although most flights still had seats available.

American Airlines' daytime flights filled up faster than nighttime flights, and the same applied to flights to New York compared to those to Washington, DC. The advance reservation rate for the inaugural Royal Coachman nonstop service from Los Angeles to Washington on June 3 was only 50%, but by the time of the flight's departure, the DC-7 had a full load of 85 passengers.

AA ORDERS MORE DC-7S

American Airlines placed an order for 19 additional Douglas DC-7s at an estimated cost of $35-40 million, scheduled for delivery between June 1957 and January 1958. This order raised the total number of DC-7s still pending to 25 and increased the company's overall orders for this type to 68.

26. American Airlines DC-7B, N344AA, c/n 45397, about to land at Los Angeles in 1962. BT by California Airmotive on Jan. 20, 1965. This propliner was CVT to a DC-7B(F) in 1960. It operated with several airlines over the years, and was purchased by Warren Cook, its last owner, on April 21, 1979. The DC-7B(F) was STD in Burbank in1996, and assumed scrapped. (Bill Larkins)

27. American DC-7, at Chicago Midway in 1957. N324AA, c/n 44145, 'Flagship Canada', DEL: June 22, 1954. BT by GALCO April 8, 1959. Sold to Lebanese International Airways in April 1963, REG: OD-AEJ. BT by ALIA Jordanian Airlines July 1, 1964, REG: JY-ACO. Destroyed: On June 5, 1967, while on the ground at Damascus-Mezzeh Airport in Syria, this aircraft was destroyed beyond repair by an Israeli air strike during the Six-Day War. (courtesy of Pat Bukiri and the Midway Airport Historians)

PAN AMERICAN WORLD AIRWAYS

EXTENDED RANGE FOR PAA DC7Cs

Pan American World Airways explored the possibility of extending the range of its Douglas DC-7Cs to provide continuous year-round service in the North Pacific. The primary obstacle on the Pacific Great Circle Route, for which PAA was an applicant, was the winter headwinds of up to 70 mph when flying westbound. The DC-7C's fuel capacity of 7,824 gal (29,617 l) and range of 4,440 miles (7,145 km) were insufficient to cross the Pacific. The DC-7C would run out of fuel if it tried to fly nonstop from Seattle to Tokyo, a distance of 4,800 miles (7,725 km), against persistent headwinds of this magnitude.

Subsequently, Douglas advised PAA to add approximately 106 gal (400 l) of fuel by utilizing the nooks and crannies in the wings. Douglas believed this would be enough to overcome the headwinds and achieve almost 100% year-round schedule reliability for PAA.

By reducing the payload to 12,500 lb. (5,670 kg), the long-range cruise with reserves could be extended to 5,100 miles (8,208 km), meeting all eastbound conditions in the North Pacific year-round and even some westbound conditions (primarily during the summer). This left only the strong winter headwinds that could impact year-round schedule reliability for nonstop operations, and further payload reductions would not overcome these unless the DC-7Cs were supplied with more fuel.

The issue was that the additional 400 gal (1,514 l) of fuel would need to be taken from the payload unless the aircraft's maximum gross takeoff weight could be increased. That was indeed feasible. Douglas had already informed airlines of a performance limit increase from 139,000 to 140,000 lb. (63,049 to 63,504 kg). The latter figure represented the structural limit of the DC-7Cs in 1956.

PAA would need to wait until 1958 for the availability of the modified DC-7Cs. However, that was far too late. By then, airlines would be firmly focused on entering the jet era, and the DC-7s would fade into obscurity.

PAA's FIRST DC-7C FLIGHT TO PARIS

Pan American's first DC-7C landed in Paris on Saturday, May 26, after a 14,800-mile (7,724 km) flight from Miami, carrying media personnel onboard. The DC-7C averaged 350 mph (563 kph) and reached a top speed of 450 mph (724 kph), aided by tailwinds. The aircraft consumed 7,135 gal (27,010 l) of fuel and had enough in reserve to land in either Frankfurt or Madrid.

PAA DC-7s TO HAWAII

Pan American began operating Douglas DC-7Cs between San Francisco and Honolulu on Thursday, July 26, and from Los Angeles to Honolulu on Tuesday, July 31. Starting Thursday, September 2, DC-7Cs were also deployed on the route between Seattle/Portland and Honolulu, with the service later expanding to other Pacific ports in the fall.

PAA also allocated 8 DC-7Cs to the Pacific-Alaska division to supplement the 16 Boeing 377 Stratocruisers serving the Pacific area and Alaska. Pan American maintained all of its Stratocruisers in service throughout 1956. PAA's new DC-7C schedules included three daily round-trip flights between San Francisco and Hawaii and two round-trip flights between Los Angeles and Hawaii.

Inaugural flights with the DC-7Cs were provided at regular pricing, labeled 'President Service' by PAA. Afterwards, the DC-7Cs operated in a dual configuration with first-class and tourist-class service. All of PAA's Stratocruisers in the Pacific region were also employed in mixed-class service.

CONTINENTAL AIRLINES

In late June 1956, Convair's new jetliner, the CV-880, was unveiled to the public. This airliner resulted from a collaborative effort involving Howard Hughes (and TWA), Delta Air Lines, Convair, and the General Electric Company. Initially named the 'Skylark 600,' the aircraft was later renamed the 'Golden Arrow.' This new designation stemmed from the plane's unmatched speed of 609 mph (980 kph) and its distinctive metal color. Delta ordered 10 units, while Howard Hughes, an enthusiastic advocate of the 'Golden Arrow,' placed an order for 30.

28. PAA DC-7C, N731PA, c/n 44873, 'Clipper Bald Eagle', DEL: Nov. 5, 1956. CVT to DC-7F in 1959. BT by Liberty Air March 15, 1965, REG: N7314. BT by British carrier Trans Meridian (TMA) Jan. 1966, REG: G-ATMF. LSD to BOAC May 18, 1967. RET to TMA June 1967. LSD to Trans African AW, June 1968. RET to TMA Jan. 1970. BT by French airline Transportes Aeriens Reunis May 1972. B/U Nice, November 1972. (The University of Miami Libraries)

The Houston-based airline Continental Air Lines (CAL) took issue with Convair, TWA, and Delta's use of the appellation 'Golden Arrow' to characterize the new Convair jetliner. CAL asserted that 'Golden Arrow' and 'Silver Arrow' were announced by the airline in early 1956 as names for its DC-7B first-class and coach flights on the company's Los Angeles-Denver-Kansas City-Chicago route, which was scheduled to begin service on May 1, 1957.

CAL intended to employ the term 'Arrow' extensively in all forms of advertising. The term would also be used on CAL's fleet of Viscount 810-40s, which were slated to be put into service on major routes in mid-1958. CAL would also offer a Chicago-west coast service with its Boeing 707 'Arrows' by August 1959, which was in advance of the start of operations of the Convair CV-880 'Golden Arrow.'

Continental Airline's complaints were largely ignored. Then, in mid-July 1956, Convair announced that it was dropping the 'Golden Arrow' moniker and renaming its jetliner the Convair 'CV-880'. That announcement would have been music to the ears of Continental's president, Robert F. Six.

BRANIFF INTERNATIONAL AIRWAYS

During the weekend of October 20 and 21, Braniff International Airways (BIA) launched its Douglas DC-7Cs into domestic service on the New York-Texas and Chicago-Texas routes. Initially, the Texan airline also planned to operate its DC-7C 'El Dorado' service on international routes to South America. However, fierce competition from airlines using DC-7Bs and Super Constellations on the Texas route required a change in its plans.

For South American routes, Braniff aimed to deploy DC-7Cs to that region in the spring of 1957, possibly as late as June 1957.

Braniff's first El Dorado services included one daily round trip flight on each of the following routes: New York-Dallas-San Antonio; New York-Washington-Nashville-Dallas-San Antonio; Chicago-Dallas; and Dallas-Houston. Nonstop flights between New York and Dallas took 4 hours and 40 minutes southbound and 4 hours and 25 minutes northbound.

The DC-7Cs of Braniff were not modified for the shorter domestic routes.

Braniff's DC-7Cs were arranged in a first-class/tourist-class configuration. First-class guests were provided with 16 seats. Four seats were positioned behind the main cabin door and buffet lounge section at the back of the aircraft. The other 12 first-class seats were between the forward lavatories and the cockpit. This setup placed the 55-seat tourist compartment between the two first-class sections.

29. Braniff DC-7C, N5903, c/n 45071, at Chicago Midway in what looks like a clear sky. DEL: Nov. 30, 1956. BT by Club America on Nov. 30, 1969. LSD by 'Grand Funk Railroad', an American rock band from Michigan. RET to Club America June 1971.LSD to Magic Carpet Air Travel Aug. 1971. NN to Traventure Air Travel 1972. BT by A. Leddya'35' Sold to T & G Aviation Inc. June 1976. Accident: On Oct. 9, 1986, the DC-7C was flying a locust spraying mission for US AID in Senegal. There were 3 crew members and one passenger onboard. Shortly after takeoff from Dakar-Yoff Airport, white smoke started to come out of engine no.3, while the fire alarm for no. 4 engine sounded. The crew shut down and feathered no.4 engine. Suddenly, the aircraft lost height and crashed, exploding into flames. The passenger was critically hurt, tragically, all three crew members perished. (courtesy of Pat Bukiri and the Midway Airport Historians)

DOUGLAS AIRLINERS OPERATED BY US SCHEDULED AIRLINES 1954-1955				
MODEL	1954 DOMESTIC	1954 INTERNATIONAL	1955 DOMESTIC	1955 INTERNATIONAL
DC-3	299	41	302	10
DC-4	109	76	89	78
DC-6	185	179	173	226
DC-7	61	41	63	53

COMMERCIAL AIRCRAFT ORDERS 1956-1958				
AIRLINE	MODEL	YEAR OF DELIVERY		
		1956	1957	1958
AMERICAN	DC-7	14	0	0
BRANIFF	DC-7C	4	3	0
CONTINENTAL	DC-7B	0	5	0
DELTA	DC-7	0	10	0
EASTERN	DC-7B	8	8	12
NATIONAL	DC-7	0	8	0
NORTHWEST	DC-7	0	8	0
PAN AM	DC-7B /7C	17	16	0
UNITED	DC-7C	2	15	0

INTERNATIONAL AIRCRAFT ORDERS 1956-1958				
AIRLINE	MODEL	YEAR OF DELIVERY		
		1956	1957	1958
AEROVIAS BRASIL	DC-7C	0	0	3
ALITALIA	DC-7C	0	0	2
BOAC	DC-7C	5	5	0
CMA - MEXICANA	DC-7C	0	4	0
JAPAN AIR LINES	DC-7C	0	0	4
PANAIR DO BRASIL	DC-7C	0	3	0
REAL (BRAZIL)	DC-7C	0	8	0
SABENA	DC-7C	2	7	0
SAS	DC-7C	8	6	0
SOUTH AFRICAN	DC-7C	3	0	0
SWISSAIR	DC-7C	2	2	0
TAI (FRENCH)	DC-7C	0	2	0

DC-7 OPERATORS 1955 (USA INTERNATIONAL)

AIRLINES	DC-7/ 7B
AMERICAN	25
DELTA	11
EASTERN	12
NATIONAL	4
PANAGRA	5
UNITED	7
TOTAL	**64**

DC-7 OPERATORS 1955 (USA DOMESTIC)

AIRLINES	DC-7
AMERICAN	25
DELTA	11
EASTERN	12
NATIONAL	4
UNITED	25
TOTAL	**77**

INTERNATIONAL AIRLINES

JAPAN AIR LINES

In April 1956, Japan Air Lines ordered 4 Douglas DC-7Cs for delivery between December 1957 and April 1958. The DC-7Cs reduced the flight time from San Francisco to Tokyo to 17 hours from 25 hours (with JAL's DC-6Bs). JAL's DC-7Cs were equipped to carry 28 first-class and 30 economy-class passengers.

A Japan Air Lines DC-7C, JA6305, c/n 45471, 'City of Hong Kong', DEL: April 7, 1958. CVT to DC-7F in 1961. BT by SAS in Sep. 1962, REG: SE-CCI, 'Eric Viking'. It operated with a few other companies. It last operator was Irish airline Aer Turas, purchased it in OCT. 1972, REG: EI-AWG, 'City of Limerick'. Accident: On Sunday March 3, 1974, the Aer Turas DC-7CF was on a charter flight from Dublin to Luton, carrying newspapers and horse transportation equipment back to Dublin. The aircraft was damaged beyond repair after overrunning the runway while landing at Luton airport. The crew failed to accomplish reverse thrust after touchdown and applied forward propulsion unintentionally during the landing roll. The ensuing overrun was made worse by the use of the emergency pneumatic brakes, which caused the bursting of the 4 main wheel tires, resulting in a loss of stopping capability. No casualties. (Template by MED. Livery by TG)

SAS

SAS was the first European airline to introduce the DC-7C into service. In September, the DC-7C made its inaugural flight from Scandinavia to New York, stopping in Prestwick and Gander. Still, the DC-7C reduced the flight duration of the DC-6B by two and a half hours. Then, in October, when nonstop service began between Scandinavia and New York, four and a half hours were saved.

The first of 14 DC-7Cs ordered by SAS arrived in Stockholm on August 10 after a nonstop flight from New York that lasted 11 hours and 48 minutes. The DC-7C was SAS's flagship until 1960 when Douglas DC-8 jetliners were introduced. The DC-7C could accommodate 89 passengers in tourist class or 40 in first class. SAS received its first three DC-7Cs by the end of September.

In February 1957, when the DC-7C was added to SAS's 'Global Express' service from Scandinavia to Tokyo, the flight duration was reduced from 53 hours (via the Near East, Pakistan, and Thailand route) to 33 hours via the Polar route. In addition, the distance between Los Angeles and Scandinavia via the trans-polar route was significantly shortened.

SAS RECORDS

On November 16, 1956, SAS set a new record for the longest continuous flight by a commercial airline by traveling 6,005 miles (9,664 km) nonstop from Long Beach, California, to Stockholm with a DC-7C. The journey was monitored by Federation Aeronautique Internationale (FAI) and averaged 277 mph (446 kph) over the 21 hr. 44 min. duration. The previous record of 5,700 miles (9,173 km) was established in 1953 by a Los Angeles to Paris flight by a DC-6B.

On another flight in early December, an SAS DC-7C completed a New York to Hamburg nonstop flight in the record time of 9 hrs. 56 mins.

30. An SAS DC-7C flying above the clouds in the 1950s. This is LN-MOD, c/n 44930, 'Guttorm Viking', DEL: Sep. 25, 1956. This aircraft only flew with SAS. WFU & B/U at Kastrup-Copenhagen, Denmark in April 1968. (Nordiska Museet/ Nordic Museum Collection. Photo by K.W. Gullers)

SWISSAIR & SABENA RECEIVE DC-7Cs

Swissair's executive committee formally welcomed the airline's first DC-7C in Los Angeles on November 2. On November 12, the aircraft was flown to Zurich and averaged 373 mph (600 kph) on its nonstop Atlantic voyage from New York. It was put into operation on the North Atlantic run on December 21.

After Swissair, it was Sabena's turn. On November 14, the Belgian airline staged a christening ceremony for the first of its 10 Douglas DC-7Cs in Santa Monica. A few days later, the aircraft was flown to Brussels for crew training and familiarization before entering service across the Atlantic on January 1. With its new DC-7Cs, Sabena enhanced its North Atlantic schedule frequency from 10 to 14 weekly flights during the 1957 travel season. The DC-7Cs also saw service on the company's Congo route.

31./32. Two DC-7s from two European national carriers. At the top is Sabena's DC-7, OO-SFK, c/n 45495 at Rome in 1958. This was Sabena's tenth and final DC-7. It was delivered to Sabena on April 28, 1958, and sold to Spanish company Spantax in June 1966. In the bottom photo is a Swissair DC-7C HB-IBL, c/n 45062, 'Genève', delivered to the airline on Dec. 5 , 1956. It was sold to Riddle Airlines on Oct. 11, 1960, and was converted to a DC-7CF in December 1960. (Top photo: Maurice Luyten, Bottom: ETH Zurich)

1957

DOUGLAS

The new and improved Douglas DC-7 entered full production in 1953. By January 1957, a total of 156 DC-7Cs, 40 DC-7Bs, and 39 DC-7As had been delivered by Douglas. Naturally, the DC-7 series benefited greatly from the development and engineering of the DC-6 program; therefore, it did not incur nearly the same development costs.

Profit margins for the DC-6 and DC-7 series airliners increased as a result. In fact, after fulfilling commercial contracts, the plane manufacturer's profit margins were about 12% before taxes. From June 1955, when Douglas joined the jet transport race, until January 1957, the company received orders for 122 DC-8s.

In January 1957, the company's commercial aircraft backlog was as follows:

MODEL	AIRCRAFT ORDERED OR UNDELIVERED	COST $
DC-6A	37	43,351,000
DC-6B	68	87,455,000
DC-7	29	55,688,000
DC-7B	70	137,964,000
DC-7C	74	171,321,000
DC-8	122	615,522,000
TOTAL	400	$ 1,111,301

After the 1956 fiscal year, the commercial backlog represented approximately half of the company's total orders, which amounted to $2.2 billion. Reported sales for the 1956 fiscal year reached a postwar high of $1,074 million, an increase from $868 million the previous year. Furthermore, in 1956, net income stood at $33.2 million, or $8.96 per share, compared to $28.2 million, or $7.65 per share, in 1955.

JUNE DELIVERIES

In June, Douglas delivered 16 DC-6B, DC-7, and DC-7C airliners from their Santa Monica facility, marking the highest rate of deliveries since the introduction of the DC-7 four years earlier. This production rate was sustained until June 1958. The following aircraft remained on order and were awaiting delivery: 81 DC-7s, 48 DC-6Bs, 42 DC-7Cs, and 34 DC-6As.

AMERICAN AIRLINES

American Airlines increased the number of cities served by its luxury 'Mercury' service from three to twelve. Previously limited to DC-7 nonstop flights between New York and Los Angeles and Washington to Los Angeles, the service was expanded on January 15 to include Chicago, San Francisco, Detroit, Boston, and Cincinnati. Furthermore, on February 3, it was added to Dallas, Fort Worth, Tucson, and Phoenix. The 'Mercury' service included deluxe meals, red carpet terminal service, reserved seats, and other features.

On June 2, American launched two nonstop DC-7 'Mercury' flights in each direction between St. Louis and New York and a nonstop DC-6 service between St. Louis and Dallas. The carrier also introduced a one-stop round trip that connected St. Louis, Cleveland, and Buffalo.

On August 1, American Airlines inaugurated nonstop service between Boston and Los Angeles with Douglas DC-7s. This service allowed travelers flying between the West Coast and Europe to connect in Boston instead of another East Coast city, resulting in significant time savings in many cases.

The airline also completed a $250,000 makeover of its passenger facilities at New York's LaGuardia Airport, which included doubling the space in the ticketing and gate areas and installing a self-claim luggage system. The facility was used until the Port Authority's airport redevelopment was completed.

33. American Airlines DC-7B, N349AA, c/n 45402, DEL: Oct. 11, 1957, seen here at Chicago Midway Airport in 1960. BT by Aero-Nord April 1, 1965, REG: OY-ANA.BT by Swedish Red Cross April 1969. Sold to French company, Air Fret, 'Ville de Montelimar', in Aug. 1970, REG: F-OCPZ. WFU & STD at Nimes, France June 1971. B/U at Nimes in 1973. (courtesy of Pat Bukiri and the Midway Airport Historians)

On September 8, American Airlines launched a new luxury service between New York and Chicago using Douglas DC-7 aircraft. The newly introduced service consisted of two daily round trips, departing from New York and Chicago at 4:00 pm and 5:00 pm, respectively.

Finally, in December, American inaugurated a nonstop service between Chicago and Mexico City on January 5. This marked the first time a US carrier offered nonstop service between the two cities. The daily service was provided by a DC-7 in a combined first-class and tourist-class configuration, accommodating 32 (first-class) 'Mercury' passengers and 38 'Royal Coachman' passengers.

CONTINENTAL AIRLINES

'CONTINENTAL CLUB COACH' – A GREAT WAY TO FLY

Continental Airlines introduced a record number of innovations in 1957 when it began operations on its new Chicago-Denver-Los Angeles route. For example, two handset clock faces were installed on the baggage carts, one showing the airplane's arrival time and the other indicating when the baggage was delivered. Passengers could buy beverages such as bourbon, Scotch, Manhattans, and Martinis for $1 a glass or a 'short' beer for 20¢.

Additionally, they could reserve seats in advance, and male passengers could make a reservation in a 'stag smoker' lounge (an informal social gathering for men) where pipe and cigar smoking was allowed. The lounge was situated at the rear of the aircraft.

Upon arrival, CAL's baggage handlers were prepared to deliver baggage in 10 minutes or less. CAL's passengers enjoyed this exceptional baggage service for the price of a coach ticket.

Instead of following the typical practice of inaugurating a new route with mixed deluxe and standard fare accommodations, CAL bypassed that step and opted for 100% coach. However, it offered a more upscale experience than a standard coach service. Even the five-abreast seating was staggered, which created less shoulder contact and allowed passengers next to each other to utilize different areas of the armrests. 'Continental Club Coach' was the term used for CAL's unique service that quickly attracted the attention of many passengers.

Here are some of the privileges CAL's coach passengers enjoyed. While boarding the aircraft, they walked on a gold carpet surrounded by gold foundations and ascended steps of metal poles painted gold under a canopy of blue fiberglass.

In the aircraft, they could buy tasty meals such as fried chicken ($1.25), sirloin steak ($2.50), and other meals for a minimal price.

Moreover, they could use an observation lounge at the rear of the aircraft with five seats and a 'picture' window for photographers.

On the ground, the passengers were transported across airport ramps in Volkswagen 'Step-Saver' coaches when connecting with other carriers. They were also transported from the airport terminal to the aircraft they were to board.

These services were offered at no extra cost. Continental quickly became the airline that many wanted to fly with. The good old days of flying were unquestionably superior to the services provided by airlines today, which are subpar at best.

PAN AMERICAN WORLD AIRWAYS

34. Continental Airlines DC-7B, N8213H, c/n 45196, DEL : Aug. 20, 1957. Seen here at Chicago Midway in 1959. WFU & STD at Los Angeles, in Aug. 1963. BT by California Airmotive March 1, 1964. BT by AREA Ecuador in May 1964, named 'Pichincha'. WFU & STD at Quito Ecuador in 1969, later B/U. (courtesy of Pat Bukiri and the Midway Airport Historians)

Pan American World Airways launched a service from New York to Nassau on Thursday, February 21. On this route, the airline used DC-7Bs exclusively to transport first-class passengers until April 28, when a mixed first-class and tourist configuration was adopted.

In March, PAWA assigned DC-7Bs to its New York-Buenos Aires service. Thus, on all five weekly first-class flights between New York and Buenos Aires, Pan Am's DC-7Bs replaced the DC-6Bs. Still, tourist flights to South America continued to be operated by DC-6Bs.

On April 28, PAWA launched a daily mixed first-class/tourist-class service between New York and Lisbon, Portugal. DC-7Cs were utilized on four flights that did not stop in the Azores for refueling. The carrier's DC-6Bs operated the subsequent three flights. The new route took passengers through Lisbon, Barcelona, Spain, Nice, France, and Rome, Italy.

On December 30, 1958, PAWA began more regular flights to San Juan, Puerto Rico, offering a new daily nonstop first-class DC-7B service and two DC-6 first-class inter-island flights.

The round-trip DC-7B flight departed New York at 9:00 am and arrived in San Juan at 3:20 pm, both local times.

NORTHWEST ORIENT AIRLINES

In February, Northwest Airlines negotiated a bank credit agreement with a maximum credit limit of $38.5 million at 4.25% APR. The funds were utilized to help finance the 14 DC-7Cs and 10 DC-6Bs on order, which totaled $55 million. NWA also had to sell four Super-G Constellations and some DC-4s to help pay for the new Douglas propliners.

On March 25, Northwest Airlines took delivery of the first of 14 DC-7Cs ordered from Douglas and promptly sent it to Seattle-Tacoma International Airport for pilot training. NWA received seven more DC-7Cs in 1957, with the remaining six arriving in early 1958.

With a weekly round trip, Northwest Airlines launched a DC-7C service between Seattle and Tokyo on April 11. On April 28, the DC-7C operation between the two locations was increased to four round trips per week.

On September 22, NWA commenced nonstop service between New York and Seattle and New York and Portland utilizing its new DC-7Cs. The flights were operated on a combination of tourist and first class. Northwest also received 7 of the 14 DC-7Cs ordered by September 1957.

35. Northwest Orient Airlines DC-7C, N292, c/n 45462, DEL : Jan. 17, 1958. Accident: On July 14 1960, NWA Flight 1-11, DC-7C, N292 took off from Okinawa-Naha Airport, Japan and was destined for Manila International Airport in the Philippines. The airliner ditched in the Pacific Ocean, around 77 miles (124 km) northeast of Manila Airport, following reports of fire in the left wing and the loss of the No. 2 propeller. Every single one of the 58 passengers and crew members managed to evacuate the aircraft safely. Forty-four sustained minor injuries, while one female passenger lost her life. The probable cause of this accident was the internal failure of No. 2 engine, resulting in oil contamination, loss of oil supply, subsequent loss of the No. 2 propeller assembly, and fire-in-flight, which necessitated a ditching. (courtesy of Pat Bukiri and the Midway Airport Historians)

IMPERIAL SERVICE

In May, Northwest Airlines introduced a new service called 'Imperial Service' for its first-class passengers. Complimentary champagne, reserved seats, and other perks were provided to passengers on transpacific DC-7C and 1049G Super Constellation flights and flights between the Twin Cities and Washington. Champagne was served before every dinner, in addition to the standard beverage service (which was charged for). Every Imperial flight was loaded with trays of hors d'oeuvres. Passengers received miniature Japanese Kokeshi dolls on flights from the Twin Cities to Washington. The new service was established to guarantee that NWA received its fair share of business on highly competitive routes.

EASTERN AIR LINES

EAL operated its initial scheduled flight to Mexico in mid-July, eleven years after being granted authorization by the United States authorities.

EAL initiated daily flights between New York and Mexico City and Mexico City and New York, with mandatory layovers in New Orleans, utilizing DC-7B aircraft. The scheduled total elapsed time for the flights was 7 hours. 50 min, including a one-hour, 15-min layover in New Orleans.

In 1946, EAL was granted US authorization for the route. However, President Truman revoked the authority, and President Eisenhower reinstated it in March 1957 following the ratification of the bilateral agreement with Mexico.

The Miami-based airline's main competitor over the route was Aeronaves de Mexico (AdM). Although AdM had the authority to fly to the US, it had no suitable airliners for operations in the United States. Thus, AdM, which aimed to commence operations by the middle of August, searched for leasable aircraft such as the DC-6Bs or Lockheed Super Constellations. In the meantime, Eastern was having a free run to Mexico with no other competitor in sight.

AdM was also authorized to operate the nonstop service New York-Washington-Mexico City, which was granted following the bilateral agreement in March. However, as the Mexican carrier awaited the delivery of two Bristol Britannia turboprop transports it had placed on order before it could start operations.

During the first month of its nonstop service from Mexico City to New Orleans, Eastern Air Lines had a 61% load factor. Eastern transported 1,289 passengers from Mexico City to New Orleans, of which 565 continued to New York.

36. Eastern Air Lines DC-7B, N844D, c/n 45450, at Chicago Midway in 1958. DEL: March 26, 1958. BT by California Airmotive July 27, 1965. Then LSD to Overseas National AW April 1966.RET to California Airmotive in 1968. BT by Allied Aircraft Dec. 11, 1970, then B/U. (courtesy of Pat Bukiri and the Midway Airport Historians)

UNITED AIR LINES

On September 15, United commenced nonstop service between New York and Seattle-Tacoma. During the journey, UAL utilized its 58-passenger DC-7s. The airline completed the eastbound journey in 7 hours and 40 minutes, while the westbound trip took 8 hours and 45 minutes. The new UAL service was made available for first-class passengers only.

On November 12, UAL launched a daily DC-7 air coach service from Boston to Chicago and Los Angeles. The new service was named 'Custom Coach.' The DC-7 departed Boston at 9:45 am EST and arrived in Los Angeles at 5:25 p.m. PST, with a stop in Chicago. The return flight left Los Angeles at 11:50 am PST and landed in Boston at 12:54 am EST, flying via Chicago and New York.

37. United AL DC-7, N63013C, c/n 44265, 'City of San Francisco'. This photo was taken on April 22, 1954, and shows one of the DC-7s just delivered to United's Western Maintenance Base at San Francisco. The other one was N6302C and neither plane had gone into service yet. It was sold to BMR Aviation on Aug. 7, 1963. WFU & STD at Ontario California on Aug. 1963. B/U Dec. 1964. (Bill Larkins)

INTERNATIONAL AIRLINES

SAS

SAS launched a scheduled passenger service on its new Polar route from Stockholm and Copenhagen to Tokyo via Anchorage, Alaska, on February 24. However, the airliners from Tokyo to Copenhagen faced heavy headwinds and had to land in Oslo, Norway, to refuel before proceeding to Copenhagen. SAS' DC-7Cs were used for the service, the outcome of long planning and expansion of the SAS Arctic service to Los Angeles via Sondre Strom fjord and Winnipeg. SAS operated the service twice a week in each direction. Scheduled flight times for the various sectors were as follows:

	EASTBOUND	WESTBOUND
Stockholm-Copenhagen	1 hr. 30 mins	1 hr. 25 mins
Copenhagen-Anchorage	16 hrs. 10 mins.	16 hrs. 5 mins
Anchorage-Tokyo	13 hrs. 25 mins.	12 hrs. 25 mins

The service had a 50-minute stop in Copenhagen and a one-hour layover in Anchorage. The arrival of SAS' exceptionally long-range DC-7Cs enabled the opening of the new route, which was the first regular passenger flight across the North Pole. Though the SAS feat in creating this new route in the 1950s, which saved around 22 hours on the voyage time to Tokyo, should not be underestimated, it is worth noting that the route taken was far from a Great Circle track between Europe and the Far East. Such a course would have involved flying over USSR territory almost the whole time, and the prospect of obtaining permission to do this from the Soviet Union during the Cold War era was impossible.

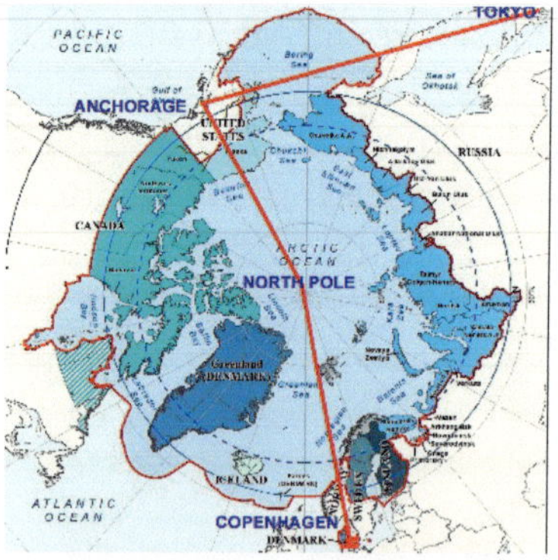

This was the route taken by SAS for its Polar service from Europe to Tokyo, Japan. (map by US Arctic Research Commission / details by author)

Thus, the best option was the revised route with its right-angled turn at Anchorage (see map).

From the polar region to Africa, East Asia, and South America, SAS has been a truly global airline, and it has been the case from the 1950s until today in 2023; that's 70 years of excellent worldwide service.

On November 21, 1957, SAS launched an initial tourist-class service between Europe and South Africa, including direct connecting flights from the United States. SAS' DC-6Bs initially operated the new service, which could carry 45 passengers: 25 in tourist configuration and 20 in sleepers and first-class dormette seats (adjustable seats that can be fully reclining). From April 1958, the Scandinavian carrier began servicing the route with the Douglas DC-7C.

38. SAS, one of the classic airlines of the 20th century, which is still flying the world's skies. This is another beautiful photo of SAS DC-7C LN-MOD taken in 1957 for the airline by K.W. Gullers. (Nordic Museum / Nordiska Museet)

SAS RECORD

The SAS DC-7C, which made the first Copenhagen-Tokyo polar flight, returned to Scandinavia via India and the Middle East. The round-the-world flight was completed in 71 hours and 6 minutes of flying time for the 18,487 miles (29,752 km).

KLM

The first of 10 DC-7Cs ordered by KLM was delivered to the Netherlands in April. Subsequently, on June 3, the type was put into operation on the North Atlantic route, specifically to and from New York; the initial crossing took 15 hrs. 36 mins.

The arrival of additional DC-7Cs saw them assigned to the Johannesburg, Montreal, and South America routes in July, August, and October, respectively.

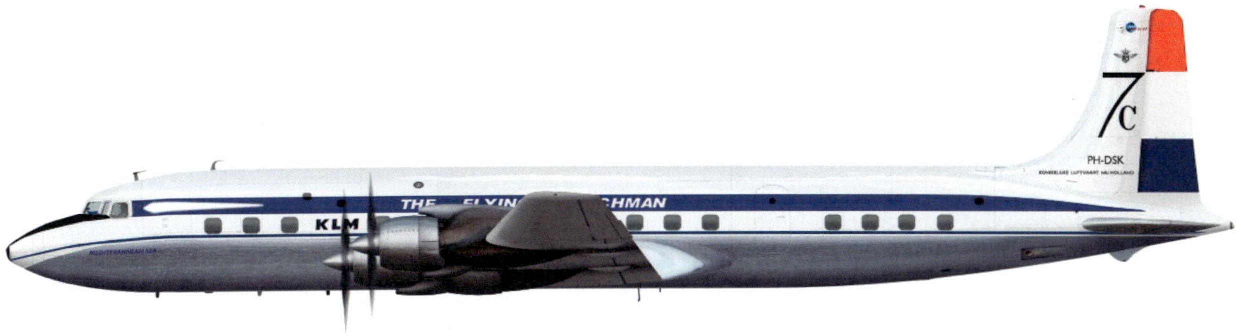

KLM DC-7C, PH DSK, c/n 45189, 'Mediterranean Sea', DEL: July 30, 1957. TFD to Martin's Air Charter Oct. 1962. BT by Modern Air Transport May 6, 1965, REG: N383M. BT by Golden Odyssey Inc. April 1968. Repossessed by Modern Air Transport Aug. 1968. WFU & STD at Fort Myers Florida, 1969. Later B/U. (Template by MED / Livery by TG)

SWISSAIR

NEW SWISSAIR ROUTES

On April 7, Swissair introduced a second service to South America (Buenos Aires) operated by its DC-6Bs. In July, DC-7Cs replaced them. Then, on May 7, the Swiss national airline launched a service between New York and Cologne/Bonn Airport, with two weekly nonstop flights using DC-7Cs on Thursdays and Saturdays.

On May 8, Swissair started a nonstop service between New York and Lisbon, providing two flights per week with its DC-7Cs. Additionally, these flights continued to Zurich and Geneva. Swissair became the first European airline to offer a service between the United States and Lisbon, Portugal. Moreover, this service was part of Swissair's summer timetable, including nonstop flights from New York to Cologne/Bonn and Switzerland. In the summer of 1957, Swissair operated nine weekly flights between Europe and New York.

Swissair introduced the DC-7C into its South American route network on July 4, 1957. Intermediate stops at Lisbon and Recife were removed, cutting flying times from Switzerland to Rio de Janeiro from 25 hours 45 minutes to 20 hours 40 minutes, to Sao Paulo from 27 hours 55 minutes to 22 hours 45 minutes, and to Buenos Aires from 33 hours 35 minutes to 27 hours 35 minutes.

On July 5, Swissair launched a second weekly DC-7C flight to Tokyo, departing from Zurich and Geneva on Fridays and flying via Athens, Beirut, Karachi, Calcutta, Bangkok, and Hong Kong. Starting July 9, the Tuesday flight to Tokyo was rerouted to include Cairo as a new point of call, along with Karachi, Bombay, Bangkok, Manila, and Tokyo. After a busy summer season, Swissair ordered an additional Douglas DC-7C for delivery in late 1958.

39. A fascinating scene on the eve of the jet era at Zurich-Kloten Airport in Switzerland in the late 1950s. A Swissair DC-7C, HB-IBL, is at the front of a pack of three Viscounts; two British European Airways, and one Alitalia, an unknown DC-4 taxiing (which might be a Seaboard & Western plane, as that cargo outfit was often seen at Zurich from the 1950s through the 1970s), and a Swissair DC-3. (ETH Zurich)

COMPANIA MEXICANA de AVIACION

Compañía Mexicana de Aviación began operating DC-7Cs on its Los Angeles-Mexico City-Havana route on March 20, replacing the DC-6s, which were shifted to CMA's domestic route network. On March 17, a pre-inauguration flight aboard a Douglas DC-7C, primarily used for crew training, carried Mexican leaders and members of the media to Los Angeles for a reception. Then, on March 19, another pre-inaugural DC-7C flight with a California press party took off from Los Angeles. CMA received delivery of two DC-7Cs and got its third at the end of March. The journey from Los Angeles to Mexico City was relatively short for the DC-7Cs. Nevertheless, CMA purchased the long-range version, hoping the Mexican airline would be granted a longer nonstop route to the U.S., such as Chicago or New York, or a transatlantic connection to Madrid.

PANAIR DO BRASIL

Early in April, Panair Do Brasil, a Brazilian airline, received the first of four Douglas DC-7Cs on order. This brought the total number of the long-range DC-7C propliners delivered by Douglas Aircraft Company to 58. Douglas had sold (or had orders for) a total of 114 DC-7Cs to 13 airlines.

40. A CMA DC-7C at Chicago Midway in June 1958, XA-LOB, c/n 45127, DEL: Feb. 14, 1957. This liner was sold to Camaero Inc. in 1958, REG: N2281. The company then LSD the airliner out to 5 different airlines from 1968 to 1964, before selling it to Connie Air Leasing Inc. in Oct. 1964. This firm sold it in turn to International Aerodyne in 1965. On Feb. 21, 1966 the firm LSD it to Saturn Airways on Feb. 21, 1966, and returned in April 1966. It was WFU & STD at San Antonio Texas. In April 1967, Spanish supplemental carrier Trabajos Aereos y Enlaces SA (TAE) BT it. REG: EC-BEN. It was WFU & STD in Madrid, Spain, before being used as a disco named 'Playa del Puig', near Valencia, Spain from 1981-84. (courtesy of Pat Bukiri and the Midway Airport Historians)

41. Initially registered by Douglas as N70C, then to Panair Do Brasil, DC-7C, PP-PDO, c/n 44872, 'Bartolomeo Bueno de Silva', DEL: June 12, 1957. Accident: On Wednesday November 1, 1961, the flight from Lisbon to Recife via Ilha do Sal saw the DC-7 reaching Recife with good night visibility. Despite the pilot reporting on wind leg, the crew seemingly made a straight entry into the base leg, flying too low and outside of the normal traffic pattern. The aircraft slammed into an 84-meter-high hill, broke up 8,924 ft. (2,720 m) short of the runway 15 threshold, and to the right of the extended centerline. Pilot error was the primary cause of the accident; specifically, incorrect distance assessment, nighttime non-standard traffic pattern flying, and a failure to notice the minimal altitude during the final approach. The incorrect labeling of obstacles at night along the approach path leading to runway 15 was also one of the contributing factors. Of the 88 occupants, 45 were killed, including 7 of the 9 crew members. (ETH Zurich)

BOAC

In 1956, BOAC DC-7Cs began operations on the North Atlantic route. As a result, several Boeing 377 Stratocruisers were converted for 40-slumberette first-class service after being removed from the North Atlantic route and repurposed for West African services. The remaining Stratocruisers were refurbished into 40-slumberette configurations and continued to operate all first-class Monarch services between London and New York.

In contrast, the DC-7Cs, configured with 42 slumberettes and 13 berths, handled the majority of flights on various North Atlantic routes and operated most of the prominent Monarch services. The 'Seven Seas' were initially designated solely for first-class services. However, in February, a mixed-class version was introduced on the route, consisting of 16 first-class/4 berths, 12 slumberettes, and 33 tourist seats. Additionally, in May, a 77-seater tourist service was launched.

The BOAC DC-7Cs began mixed-class service flights across the United States to San Francisco in April, connecting with Qantas Pacific services. In April, the BOAC DC-7Cs operated flights across the US to San Francisco with a mixed-class setup, linking with Qantas Pacific services, and eventually replaced the Stratocruisers on the mixed-class London-Prestwick-Detroit-Chicago route with an all-tourist service introduced later. Additionally, that month, the DC-7C was used for an all-tourist London-Manchester-Boston-New York service. Furthermore, the 'DC-7C' operated mixed-class services to Montreal in May.

RECORD NEW YORK-LONDON FLIGHTS

BOAC flew a record 48 flights across the North Atlantic in the summer of 1957, including 38 between New York and London. The latter comprised daily deluxe-class Stratocruiser trips and 5 nonstop DC-7C tourist-class services each week in both directions.

In addition, on this route, 12 services stopped in Glasgow, 8 in Manchester, and 8 in Boston. These were mixed and tourist-class flights operated by DC-7Cs. There were also two weekly mixed-class DC-7C flights between London and San Francisco, with a stopover in New York, and six weekly flights between Chicago-Detroit and Glasgow-London (one mixed-class and two tourist-class). BOAC also scheduled 17 weekly flights between Canada and the United Kingdom over the summer.

1958

DOUGLAS

Sales for Douglas reached $1.1 billion in 1957, with military orders accounting for 69% of the total. By the end of the year, the value of outstanding orders totaled $1.7 billion. In 1957, sales of civil aircraft amounted to $344 million. A yearly high of 126 DC-6 and DC-7 airliners were delivered, and 134 DC-8 jetliners were ordered.

Douglas Aircraft Co. delivered the 1,000th plane of its DC-6/7 series in June, a DC-7 for United Airlines. June also marked the delivery of the 1,000th airliner of the DC-6/7 series at the Santa Monica Division of Douglas Aircraft Company. Explicitly produced for United Air Lines, the DC-7 was the 151st airliner of the DC-6/7 series and the 57th propliner of its kind delivered to the airline.

A total of 1,000 transports were sold to 53 airlines, comprising 23 domestic and 30 international, and the USAF and US Navy. Of this total, 176 were DC-6s, 60 DC-6As, 268 DC-6Bs, 106 DC-7s, 111 DC-7Bs, 114 DC-7Cs, and 167 were military variants. In 1947, the first DC-6 was sold for approximately $640,000, while in 1958, the more luxurious DC-6B and DC-7C were valued at $1,450,000 and $2,300,000, respectively. The approximate total value of the 1,000 aircraft was $1.3 billion. The backlog of undelivered DC-6/7 series liners totaled 41, with production continuing through November at Santa Monica.

NUMBER OF DC-7 SERIES AIRLINERS IN SERVICE AT THE END OF 1958

US AIRLINES	DC-7	DC-7B	DC-7C	INTERN. AIRLINES	DC-7B	DC-7C
AMERICAN	58	-	-	MEXICANA	-	4
BRANIFF	-	-	7	ALITALIA	-	6
CONTINENTAL	-	5	-	KLM	-	15
DELTA	17	4	-	SABENA	-	10
EASTERN	-	28	-	SAS	-	14
NATIONAL	4	4	-	SWISSAIR	-	5
NORTHWEST	-	-	15	TAI	-	3
PANAGRA	-	6	-	BOAC	-	10
PAN AM	-	7	25	SAA	5	-
UNITED	55	-	-	JAL	-	4
TOTAL		235			76	

THE LAST DC-7

In August, the last airplane on the DC-6/7 line at Douglas was No. 1041, a Douglas DC-7C for KLM. Consequently, the Dutch carrier operated the last piston-powered aircraft to be built (but not delivered).

Midway through 1958, Lockheed's Constellation production line had ceased operations, albeit with a few remaining deliveries. Meanwhile, Convair possessed a limited number of 440s that were finished but remained unsold.

Furthermore, one Douglas client was not as prompt in making the final payment as KLM was with its DC-7C. Aircraft No.1040, the aircraft directly ahead of the KLM 'Seven Seas' in the Santa Monica line, was the final DC-6B to be delivered and, as a result of the airline's late payment, it beat the DC-7C as the last propliner to be delivered. The airline was the Yugoslavian carrier JAT.

THE AIRLINES DUMP PISTON PLANES FOR JETLINERS

The airlines and aircraft manufacturers were concerned that the used-plane crisis would deteriorate during the ensuing two or three years (1958-1961) when the airlines would dump their excess propliners onto the market to make room for over 400 turbine-engine airliners.

The secondhand aircraft market for long-range, piston-engine aircraft like the Douglas DC-7 and Lockheed 1049 series was almost non-existent. With little interest in piston-engine airliners, potential buyers with enough cash or credit shopped in the turboprop or turbojet markets.

Moreover, manufacturers refused to accept trade-ins on new jet aircraft sales. However, the plane makers gave the airlines some optimism by stating that they could be obliged to use a trade-in to maintain a viable sales volume.

The airline industry was confronted with the issue of equipment obsolescence for the first time in its history. Unless a market for larger piston engine aircraft somehow sprung up, the factor of obsolescence would force airlines to keep their surplus aircraft in mothballs rather than flooding their routes with an excess of available seat miles.

Braniff was among the first carriers to mothball 2 of its L1049 Constellations, while Capital Airlines did the same for 6 of its L-1049s. By doing so, both airlines avoided incurring losses if their aircraft had remained in service.

THE DC-7 MARKET

American Airlines projected in a letter to the Civil Aeronautics Board (CAA) that the market price for the DC-7 would eventually fall as low as $200,000.

In late 1958, US airlines had 215 DC-7s either in service or on order, with 69 of the same type in service by foreign carriers. The effect of the DC-7 on the secondhand aircraft market was illustrated by United Air Lines' plans to retire 36 of its DC-7s in 1961 and American's goal of depreciating its 60 DC-7s by 1961

American stated that when it bought the DC-7s, the turbojets were not expected to be ready as early as 1958. AA also stated that it had received no proposals to purchase the DC-7s it had already listed for sale.

RETIREMENT OF THE DC-7 FLEET

American and United were the two major US airlines equipped with the most significant number of DC-7s in 1958. American Airlines operated a fleet of 58 DC-7 series propliners and had 4 DC-7s for sale in 1958, along with another 25 liners by July 1959. They asked $1.2 million for an aircraft originally costing $2 million.

However, due to the rapid turnover of piston engine aircraft required for the introduction of jetliners, the carrier declared that a fixed depreciation formula could not bind it. American estimated that a seven-year service life for its piston engine aircraft would have increased its retirement loss exposure by over $178 million.

To support its position that managerial decision-making should determine depreciation rates, AA disclosed that the initial residual value and service life of its fleet of DC-6Bs were 10% and 5 years, respectively. The depreciation life was extended to 7 years in 1954 after firsthand observations showed that the aircraft could sustain regular operations longer than initially anticipated.

Additionally, the 25 DC-7s that American acquired between 1953 and 1954 were depreciated over 7 years. In contrast, a four-to-five-year service life was applied to the 33 DC-7s delivered between 1956 and 1958 to establish a common retirement date of 1961, the year selected for the retirement of the DC-7 fleet.

UNITED AIR LINES RETIRES PROPLINERS BY 1961

In 1958, United Air Lines had 55 DC-7 series aircraft in its fleet, ranking second to American Airlines. The carrier received its last piston-engine aircraft, a DC-6A cargo liner, which brought its total fleet to 198 piston-engine planes. The conversion to a turbine jet fleet was set to begin in 1959, starting with the delivery of the first of 40 Douglas DC-8s, followed by the first of 11 Boeing 720s in 1960. Like American, UAL planned to retire its fleet of DC-7s in 1961.

UNITED MAKES A DEAL FOR DC-8 JETLINERS

In September, United Airlines and Douglas sought CAB permission for a joint sales agreement allowing the airline to trade in 10 DC-7 propliners in service to acquire 10 DC-8 jetliners, which were already on order.

The application, filed with the CAB in early September, contained separate agreements reached on October 24 between the airline and the Douglas Company for selling piston-engine propliners and acquiring jet transports.

To purchase the new planes, delivery dates and payments for the secondhand aircraft were set. The resale price of the DC-7s was fixed at $1.25 million per plane, with a $5,000 price reduction for each month the airliner was operated after its original delivery date to United Air Lines.

NWA-THE LAST CARRIER TO ORDER JETLINERS

Only one of the leading US domestic and international trunk carriers, Northwest Airlines, had not ordered any turbine airliners by April 1958. For two and a half years, NWA had a practice of keeping American aircraft manufacturers waiting while it unhurriedly contemplated the qualities of the enormous number of airliners it was being courted to acquire.

Finally, the aircraft manufacturers got a break; NWA purchased 5 Douglas DC-8s and 10 Lockheed Electras. Despite the jet competition it would face in the near future, Northwest's choice to select turboprops for its medium-range operations was particularly intriguing.

In 1958, NWA's aircraft fleet included 3 DC-7Cs, 4 L-1049Gs, 9 Stratocruisers, 16 DC-6Bs, and 18 DC-4s.

It was known that as part of the DC-8 arrangement with Northwest, Douglas agreed to buy back existing piston-engine transports for the first time in its history, which was what Lockheed had already been doing.

Lockheed struck a similar deal with Northwest. The manufacturer would accept the 9 Stratocruisers in NWA's fleet as payment for the 10 L-188 Electras.

The details of the two deals were as follows:

DC8: Deliveries would take place between March and September of 1960. One DC-8 cost $5,780,000, while the overall cost of the 5 airliners, including engines and spares, was $37,845,000. In exchange, Douglas bought 5 NWA DC-7s for $1.35 million each, for a total of $6,731,000, including related inventory.

L-188 Electra: Deliveries were made between July and December 1959. One Electra cost $2.4 million, and the total outlay was $29,654,000. Lockheed, like Douglas, established trade-in arrangements for Northwest's 9 Boeing 377 Stratocruisers, which were acquired back for $390,000 each or $3.5 million for the lot.

42. NWA DC-7C, N294, c/n 45464, DEL: Feb. 26, 1958. BT back by Douglas in 1960, then LSD to NWA in same year. RET to Douglas June 1961. BT by Riddle Airlines July 5, 1961. CVT to DC-7CF Oct. 1966. Riddle changed name to Airlift International on Nov. 29, 1963. WFU & STD in Miami, Florida, 1968. Later B/U. The NWA DC-7C is seen here smoking up the engines at Chicago Midway in 1960. (courtesy of Pat Bukiri and the Midway Airport Historians)

NATIONAL AIRLINES

National Airlines took an unusual approach to finance its jetliner procurement program. With the exception of six Convair CV-440s, it put its entire fleet of 44 piston-engine propliners up for sale. This list included 4 DC-6s, 8 DC-6Bs, 4 DC-7s, 4 DC-7Bs, 4 L-1049Hs, 8 Lockheed Lodestars, and 12 Convair CV-340s. That was one way of doing it.

MEXICO OPENS UP

A pivotal event in 1957 was the signing of a provisional bilateral agreement between the United States and Mexico in March. This agreement authorized seven routes until June 1959.

As a result of the agreement and subsequent CAB judgments, American Airlines began operating nonstop flights between Mexico City and Chicago.

Eastern Air Lines also offered nonstop flights from Mexico City to New Orleans and New York, while Western Air Lines operated nonstop from Los Angeles to Mexico City. On July 23, Eastern flew first-class DC-7Bs from New Orleans to Mexico City and from New York on September 20. By December, both routes had accumulated a total of 16,468 passengers.

American, which had previously operated between Chicago and Mexico City via Fort Worth or Dallas, launched nonstop DC-7 service on January 5, competing with Mexicana for tourist and first-class traffic. Meanwhile, Pan American lost the New York-Mexico City route to Eastern and was compelled to sell its investment in Aeronaves de Mexico (PAA was AdM's parent company).

BRANIFF INTERNATIONAL AIRWAYS

Braniff International Airways increased its revenue passenger miles on international routes by 33% in 1957 and its available seat miles in Latin American service by 21%. The airline introduced the DC-7C to its global network throughout the year, with services connecting the United States to Bogotá, Colombia. Braniff operated DC-7 flights from New York and Washington to Panama, Peru, and Brazil through Miami via an interchange with Eastern.

Additionally, in 1958, Braniff launched its most extensive tour program to Latin America, consisting of over 50 packages. The company hired catering and passenger service managers to oversee operations in various locations across the South American continent.

INTERNATIONAL AIRLINES

KLM

On November 1, KLM Royal Dutch Airlines commenced operations on a new Polar route that connected Biak, in Netherlands New Guinea, with Amsterdam and Tokyo. KLM DC-7Cs operated the new service and flew twice weekly in both directions. An intermediate stop was made in Anchorage, Alaska; however, an additional layover between Amsterdam and Anchorage was possible, dependent on the prevailing wind currents.

The Swedish authorities authorized KLM to conduct technical landings at Lulea-Kalix Airport, located about 4.5 miles (7 km) south-southeast of Lulea, Sweden, and to operate flights over specific areas that were previously inaccessible to the airline. Travel time from Amsterdam to Tokyo via the new route was approximately 32 hours, as opposed to the previous route via Karachi and Bangkok, which took 48 hours.

ALITALIA

The first of 6 DC-7Cs ordered by Alitalia commenced service between Rome and New York via Paris in January 1958. The DC-7Cs were progressively integrated into the company's North Atlantic luxury service flights, performing together with 8 DC-6Bs that were dedicated to transporting tourists.

Alitalia inaugurated a new service connecting New York to Karachi and Bombay (Mumbai) via Rome on November 1, 1958. One of the new flights served the Rome-Athens-Karachi-Bombay route, while the other served the Rome-Teheran-Karachi-Bombay line. Both flights were operated twice a week.

Furthermore, the DC-7Cs operated the vital New York-Rome route, while Alitalia's Douglas DC-6Bs flew between Rome and Asian cities. Alitalia's DC-7Cs were configured in a first/tourist/economy class arrangement.

ECONOMY CLASS TRANSATLANTIC FLIGHTS

In 1957, 1,003,000 travelers flew across the Atlantic. The airlines hoped to exceed that amount in 1958, and they did. During the 1958 summer season, the airlines provided 18,000 economy-class seats per week on transatlantic services. In all, around 200,000 extra seats were made available to passengers throughout 1958, a 20% increase from 1957.

This section focuses on the airlines that utilized DC-7s to cross the Atlantic, as well as the configuration of each airliner and the number of weekly flights.

ALITALIA: The Italian national airline scheduled 9 weekly flights during the peak summer season. Its DC-7Cs were outfitted with 16 first (rear) seats, 6 tourist (front), and 55 economy-class seats. Alitalia's highest weekly total was 495 economy seats.

BOAC: During the summer, BOAC operated 21 weekly flights between London and New York using DC-7Cs, with 16 first-class (rear), 17 tourist (main), and 37 economy seats. Flights to Canada: There were 5 weekly trips from London to Montreal by DC-7C, with 16 first-class seats (back cabin, with 4 bunks), 18 tourist seats (rear), and 63 economy seats (main and front). Britannia 312 turboprops with 12 deluxe, 16 first (rear), 15 tourist (main), and 54 economy seats operated on the London-Montreal-Detroit-Chicago line 7 times a week. There were 2 weekly trips from London to New York and San Francisco, also by Britannia 312 with 12 deluxe (rear), 27 tourist (main), and 45 economy seats. BOAC's greatest weekly total was 1,560 economy class seats.

KLM: KLM operated 14 weekly flights on their L-1049G Super Constellations, with 95-98 economy seats. Six weekly flights (to Montreal) by DC-7C with 12 deluxe (rear), 8 first-class, 35 economy (main), and 6 economy (front) seats. KLM's overall total per week was 1,576 economy seats.

PAN AMERICAN: Pan American allocated its total capacity as follows: 5% for deluxe, 23% for first, 9% for tourist, and 63% for economy. Pan Am DC-7Cs operated 12 services per week with first and economy seats, including 7 on the Arctic route to the West Coast; DC-7Cs provided 14 services per week with tourist and economy seats; DC6Bs operated 7 services per week with first and economy seats; and DC-6Bs flew 14 services per week with all economy seats. PAA's highest weekly total of economy seats was 3,500.

SAS: During the height of the summer season, SAS operated 12 weekly flights with all-economy DC-6B seating for 81 passengers, 4 weekly flights with all-economy DC-7C seating for 85 passengers, and 14 weekly flights with DC-7C featuring 28 first (rear) and 34 economy seats. The weekly peak total for economy seats at SAS was 1,788.

SABENA: Sabena provided 14 weekly flights on DC-7Cs, which featured 16 first-class, 10 tourist, and 48 economy seats, in addition to 7 all-economy flights on L-1049G Super Constellation, which accommodated 97 passengers. The highest weekly total for economy seats on Sabena flights was 1,367.

SWISSAIR: Swissair operated 8 DC-7C flights weekly, with 14 first (rear), 4 tourist, and 57 economy seats. It also offered 3 weekly flights by DC-6B, with 16 first-class and 52 economy seats. Swissair's peak weekly economy seat total was 612.

An Alitalia DC-7C, I-DUVB, c/n 45542, DEL: July 18, 1958. Alitalia was one of the last international carriers to receive the DC-7C. Its first of five DC-7Cs was delivered in January 1958. LSD by Societa Aerea Mediterranea (SAM) in 1961. BT by US company, International Aerodyne, on April 28, 1966, REG: N6000V. Sold to Voyager 1000 in May 1966. Then to Jespan Inc., in Feb. 1975. Finally, BT by T & G Aviation, July 1983. STD at Chandler, Arizona, derelict in FEB. 1997, eventually B/U. (Template by MED. Livery by TG)

THE STORY OF AN AIRLINE – TAI

The French airline Transports Aeriens Intercontinental (TAI) was founded in the immediate post-World War II years, just as the commercial air transport industry began to impact aviation history. It was founded on June 1, 1946, by Colonel Genain as the successor to the prewar (French Airline) Air Afrique, of which he was managing director.

Operations commenced on July 1, 1946, to North Africa aboard French-built Junkers Ju-52/3ms propliners. These were decorated with the new TAI livery, which featured Air Afrique's winged griffin logo and 3 Bristol 170 Freighters. In 1947, TAI switched to Douglas DC-4s.

Thanks to the DC-4s, long-distance passenger and freight routes were established to numerous locations throughout the French Empire, including Dakar, Senegal, Saigon, Vietnam, Tananarive, Madagascar, and Casablanca, Morocco.

TAI went from strength to strength. The company increased its annual financial turnover by 129% over ten years, from 1946 to 1956, an accomplishment that any company director would be proud of.

When TAI's directors issued shares on the Paris Stock Exchange, they witnessed France's only airline company's stock rise, from a nominal issue rate of 2,500 francs per share to something like 8,000 francs per share in less than two years!

In 1954, after seven years in service, the DC-4s were replaced on long-haul routes by DC-6Bs, which were great workhorses for TAI until 1958, when the carrier received DC-7Cs. The fleet in 1958 stood as follows: 3 DC-7Cs, 4 DC-6Bs, 1 DC-6, 2 DC-4s, and 4 DC-3s. In line with the times, TAI also ordered 2 Douglas DC-8s for delivery in early 1960.

TAI's management was not content to sit back and take it easy from the start but instead made every effort to improve its services and expand its areas of activity whenever favorable prospects came. A policy like this finally paid off. TAI's total flying hours were 2,180 in 1946, but by 1956, they had climbed to 26,578 hours.

TAI was also making news for other reasons. For three years, from May 29, 1953, TAI held the World record for the greatest distance flown nonstop by a commercial airline. On that May day, a TAI DC-6B proudly showing the green-and-white flying griffin logo took off from Los Angeles for the 5,985-mile (9,632-km) flight. The DC-6B landed at Orly Airport in Paris exactly after 20 hrs. 30 mins. The tired but ecstatic TAI crew achieved their goal by flying the Polar route.

TAI also became a partner in the newly founded Société Auxiliaire de Gérance et de Transports Aériens (SAGETA) in 1953. TAI personnel flew 7 SNCASE S.E.2010 Armagnac French four-engine transports from Toulouse to Saigon via Beirut, Karachi, and Calcutta on special resupply missions to Vietnam. Meanwhile, as the DC-6Bs began to replace the DC-4s in 1954, the SAGETA effort was terminated in 1955 after the French-Vietnamese agreement for France to withdraw from Indochina. (The withdrawal of French military forces was due to the disastrous French loss during the First Indochina War between France and Việt Minh in 1954. Another country would soon take over the reins in the Second Indochina War – the USA).

The French government presented TAI with an excellent opportunity two years after the SAGETA project failed. On January 1, 1956, the French government reorganized its international air network. Air France discontinued operations to Australia and the South Pacific, handing over the routes to TAI, which retired from its Africa services.

Colonel Genain's airline, TAI, always had an inclination for long-distance flights, so it was ready to take on the South Pacific. TAI's flights to the South Pacific were handled by a TAI DC-6B departing from Paris once a week.

The itinerary crossed Europe, stopping in Athens, then to Karachi, Saigon, Darwin, and Brisbane, Australia, before traversing the immense Pacific Ocean to Noumea, New Caledonia, and finally, after another ocean crossing, ultimately arriving at Auckland Airport, New Zealand.

This scenic route brought together seven different nations and seven distinct cultures and was the world's second-longest route to be operated without changing airplanes. In fact, the longest route was the other TAI South Pacific service, which flew the same line as far as Noumea before branching off to Boro Boro, the nearest airport to Papeete, Tahiti. When the passengers arrived at Boro Boro, they boarded a Sunderland flying boat that took them to their destination, Papeete.

TAI was also among the world leaders in its approach to the technical aspect of operations. Consequently, in 1955, TAI became the World's record holder for daily utilization of the DC-6B, a performance for which it was awarded the Douglas Aircraft Company's Medal of Honor.

In 1956, TAI managed to better its own figures and achieved a total utilization figure of 11 hours 33 minutes average daily usage per aircraft.

The standard of maintenance of TAI's workshop was of a very high standard in the past, and it was no surprise to those who knew of the French airline's high maintenance levels when, in 1955, the US Civil Aeronautics Administration certified the TAI workshop as their official maintenance station at Orly. Furthermore, TAI was also entrusted with maintaining all MATS (Military Air Transport Service) aircraft calling at Orly that year.

By a special arrangement with the French Government and the other two leading French international airlines, TAI was named the official French flag carrier operating in the area, which covered the whole of the South Pacific.

TAI claimed numerous exceptional flights among its accomplishments, not the least of which was the charter flight that transported more than 300 athletes to the Olympic Games in Melbourne in 1956. TAI also had the honor of transporting the one hundred thousandth Hungarian immigrant from Austria sent by the Intergovernmental Committee for European Migration in early 1957. Additionally, the yearly Islamic pilgrimage to Mecca was a magnificent event that the TAI aircrew looked forward to. TAI transported pilgrims to Jeddah, Saudi Arabia, 28 miles (45 km) from the Holy City of Mecca, every year since 1947. On the other hand, one DC-4 was turned into a nursery garden of rubber plants between Saigon, Vietnam, and Abidjan, Ivory Coast. At the same time, another was transformed into a farm when it transported pigs from Paris to Saigon.

TAI received 2 Douglas DC-7Cs from Douglas in 1957, boosting the French company's prestige and opening up new routes. Tragically, on September 24, 1959, immediately after taking off from Bordeaux for the Ivory Coast, a TAI DC-7C carrying 65 passengers crashed into a forest on takeoff. The entire flight crew and 45 of the 56 passengers on board were killed, with the other 11 passengers being critically injured.

TAI signed an agreement with Air France on May 6, 1960, to fly the South Pacific portion of the French national carrier's 'around-the-world' service via New Caledonia and Tahiti.

TAI acquired its first Douglas DC-8-32 on July 31, 1960. Following that, on September 4, the French airline launched a service using DC-8s from Paris to Darwin, Australia, via Athens, Teheran, Karachi, Bangkok, Phnom Penh, and Darwin. The Douglas jetliner then inaugurated a route from Sydney to Los Angeles via Noumea, Nandi, Papeete, and Honolulu on September 12.

On May 6, 1961, Douglas DC-8 flights were launched to South Pacific destinations. The airline agreed in principle to a possible merger with Union Aeromaritime de Transport (UAT) on September 14.

Meanwhile, on October 20, French airline 'Madair' (Madagascar Airlines) leased a DC-7C to start Tananarive (Madagascar) to Paris services.

Then, after 15 years of operations, TAI would be no more. The winged griffin silently vanished into the night.

TAI and UAT agreed to merge, and their fleets and schedules became integrated in 1962. The amalgamation of the two carriers was officially announced on October 1, 1963, resulting in a new airline, Union de Transports Aeriens, or UTA, which lasted until 1992 before being swallowed up by Air France.

TAI DC-7C, F-BIAP, c/n 45366, DEL: Nov. 9, 1957. The French airline TAI was one of the best in maintenance of its airliners, so much so, that the Military Air Transport Service (MATS) signed an agreement with the airline to service the MATS aircraft. Accident: This was the ill-fated DC-7C that crashed at Bordeaux. TAI flight 307 was a regular passenger service between Paris-Orly to Bordeaux, Mali, and Abidjan, Ivory Coast. Shortly after taking off at night (10:23 pm) from runway 23 at Bordeaux-Mérignac Airport, the aircraft took a straight-in route but failed to reach enough altitude. The four-engine aircraft impacted pine trees, stalled, and crashed in flames in a forested area 3,445 ft (1,050 m) past the runway end after flying a short course at an altitude of roughly 100 feet (30 m). The aircraft was destroyed by impact forces and a post-crash fire, killing 9 crew members and 45 passengers. The 11 surviving passengers suffered serious injuries. Cause, pilot error. The recreated flight demonstrated that a minor increase in speed during the initial section of climb-out and during a very brief critical period resulted in a significant drop in rate of climb or even a slight loss of altitude. This was due to the quick sequence of cockpit operations during this phase, as well as the rapid variation in flight parameters, and the lack of precision (even inaccuracy) of readings of certain instruments, as well as the lack of time reference and external visual references, a pilot may follow a line of flight that will drop the aircraft back near the ground if optimum climbing speed is not maintained and the altimeter is not carefully watched during this period. (Model: California Classic Propliners, Livery: TG)

THE END OF THE PROPLINER ERA

The day that ushered the end of the propeller aircraft era in the US was October 26, 1958. That day, a Pan American Airways Boeing 707-120 created history when it took off for the inaugural flight to Paris, making it the first time a jetliner would cross the Atlantic Ocean on an operational flight.

The De Havilland Comet, the world's first jet airliner, inaugurated the world's first commercial jet service on May 2, 1952. Still, the American 707 was the one that was the center of attention due mainly to the misfortune of the Comet, which suffered many accidents and lost lives.

Speaking of misfortune, the Douglas DC-7, an excellent example of what a propeller airliner should be, never reached its full potential as the jetliners stole its thunder. Undoubtedly, America in the late 1950s was ready to enter the jet era. For the DC-7, it was the end. As they say, 'The King is dead, long live the King.'

43. Established in 1962, Aer Turas Teoranta was an independent Irish airline providing charter services for passengers and cargo out of Dublin Airport. Operations started in 1963 but were put on hold in 1964 for organizational restructuring. The services were restarted in May 1965. The airline first operated the DC-7 when it purchased 4 ex-KLM Douglas DC-7CFs, and painting them with a new yellow/green livery (see photo), which was based on the multi-bar tail livery of KLM at the time. Two of the DC-7CFs were immediately resold. The 2 DC-7CFs that remained, 'City of Dublin' and 'City of Limerick,' were placed in service from Shannon, Ireland, flying racehorses and other cargo to destinations in the UK, Italy, France, and Ireland. The DC7CF in this photo is 'City of Dublin' REG: EI-ATT. In 1970, DC-7CF all-cargo charters were operated to New York and Caracas. The 'City of Limerick' was sold in April 1970, and so was the 'City of Dublin' in August 1971. As for Aer Turas, it was forced to cease operations on July 2, 2003, and file for receivership due to financial difficulties (RuthAS)

44. On November 30, 1965, Internord Aviation A/S was created as an equal commercial partnership between the Danish charter operator Aero-Nord Inc. and Ole Swedish charter airline Ostermanair Charter AB to fly inclusive tour flights, mostly between Scandinavia and the Mediterranean. The merger partners' inherited and leased 4 Douglas DC-7s and 4 DC-7Bs with both Danish and Swedish registrations. However, unable to maintain economic viability, Internord shut its doors on October 30, 1968. This photo is of Internord DC-7, SE-CNF. (Lars Söderström)

45. Vance International Airways Inc. was founded by Vance Roberts in Seattle, Washington. It began operations in 1949 and was granted interim operating authority in 1962 by the CAB as an air taxi operator and supplemental carrier. Permanent authority was granted in 1966, and Vance was awarded traffic rights to Canada and Mexico on December 15. However, the carrier's financial situation was disturbing. Subsequently, Roberts sold his airline to McCulloch of Long Beach, California in 1971, and was renamed 'McCulloch International Airlines'. At the time of the sale, Vance was operating one DC-7 and a DC-7B. The photo shows a Vance International, DC-7B, N4882C which was delivered to the company in Nov. 1968. (RuthAS)

46. Trabajos Aereos Y Enlaces (TAE) was a Spanish airline founded in 1957 by shipping magnate Naveira Aznar. However, due to difficulties in obtaining an air operator's certificate, TAE's operations were delayed until April 1967. That year, charter and all-inclusive tour services were introduced. TAE operated two DC-7Cs, and leased a BAC 1-11-402 from the manufacturer on March 1, 1969 until 1970, that was when the airline ceased operations and the jetliner was repossessed. Following lengthy refinancing discussions by the owner, Naveira Aznar, TAE was reformed and began operations once again on April 1, 1973, with the support of a technical assistance agreement with the French airline UTA (from which TAE leased 2 DC-8s). The airline was on the razor's edge in the early 1980s. TAE faced stiff competition from other charter operators, which operated more modern jet airliners compared to older DC-8s and Caravelles of TAE. Moreover, higher fuel bills and a loss of traffic due to the world recession at the time, caused the carrier to cease operations at the beginning of 1982. (Ken Fielding)

47. A BOAC DC-7CF REG: G-AOII, at London Airport on April 3, 1964. BOAC turned the DC-7Cs to cargo planes after first using them to cross the Atlantic as passenger liners. This particular aircraft was delivered to BOAC on Jan. 29, 1957. In December 1960, it was converted to a freighter. (Ken Fielding)

48. A Trabajos Aereos del Sahara (TASSA) DC-7, EC-AVP, originally operated by United Air Lines and purchased by TASSA from California Airmotive in May 1964 , is seen here stored at Palma de Mallorca on Sep. 11, 1965 (It was eventually B/U in May 1970). TASSA was a Spanish charter airline formed in 1961 primarily to undertake oil-drilling support operations in the Spanish Sahara region. Domestic Spanish (Inclusive Tour) IT work was undertaken from 1962 and international IT operations started in 1963. In May 1964, the company received its first of two Douglas DC-7s, thus substantially augmenting its transportation capabilities. However, a deficit persisted in flight operations. On October 20, 1964, the British Ministry of Transport momentarily detained a TASSA aircraft at London-Gatwick for non-payment of landing fees. Despite its precarious financial state, TASSA purchased another Douglas DC-7 in April, and in June 1965 it became entirely indebted. That month, shortly after receiving the last aircraft, TASSA declared bankruptcy. (The photo above is the DC-7 that TASSA received in May 1965 before going broke. The company never got a chance to fly it). (Ken Fielding)

49. Conair of Scandinavia was founded in October 1964 in Copenhagen, by the travel company 'Simon Spies Holdings' to provide passenger charter and inclusive-tour (IT) flights to vacation spots in southern Europe, the Mediterranean, and Africa. Between 1966–1968, the fleet was upgraded by the addition of 3 DC-7Cs. In 1970, passenger boardings reached 138,768. Throughout the 1970s and 1980s, the airline was quite popular with the Scandinavian passengers. Conair made steady growth during those years, and in 1992, passenger boardings increased to 1,152,600. In September, the parent company' Simon Spies Holdings' signed an agreement with SAS Leisure, a subsidiary of SAS (Scandinavian Airlines System), to merge the company with Scanair to form the larger Premiair. The photo shows a Conair DC-7, REG: OY-DFR, BT by the airline in Jan. 1967, and was WFU & STD in Dec. 1971. It was later used as a discotheque, before been B/U. (Lars Söderström)

50. Although there was no official DC-7 Cargo liner produced by Douglas, like there was for the DC-6A, between 1959 and 1960, United modified 6 DC-7s to DC-7BF standard, redesignating them DC-7As. This DC-7A, N6342C, taxiing at Chicago Midway Airport, was originally a DC-7, delivered to UAL on July 20, 1957, and converted to a freighter in 1959. On Aug. 31, 1966 it was bought by Airlift International. WFU& STD at Sebring Florida in 1972, later B/U. (courtesy of Pat Bukiri and the Midway Airport Historians)

51. Ostermanair Sweden was a company formed in 1962 with the name 'Svea-Flyg' by entrepreneur Göte Rosén, which operated charter and inclusive tours for Swedish tour operators to the Mediterranean region with a leased DC-6B from Hans Osterman AB. John Lennart Osterman held the majority stake in Svea-Flyg, and in 1964 Osterman paid off the other shareholders and took over the company. Previously, in the summer of 1963, Lennart Osterman had acquired 3 Douglas DC-7s from Overseas National Airways (ONA) of the US. At the beginning of 1965 the company got two additional Douglas DC-7s and put them into service. With 5 DC-7s in the fleet the company sought to collaborate with another outfit, Danish Aero-Nord, in order to operate the aircraft more efficiently. A merger was the result and on Jan. 1 1966 the new airline, named Internord Aviation A/S, began operations. Above is an Ostermanair DC-7, SE-CNH. (Lars Söderström)

52. Spantax Transportes Aereos / Spanish Air Taxis (SAT) was formed at Madrid on October 6, 1959. SAT was approved to charter larger aircraft for oil industry passenger flights in April 1960. Two Douglas DC-3s were purchased and charter services commenced from Las Palmas on December 20 to facilitate Spanish Sahara fuel resource development. In 1963/64, DC-6s and DC-7Cs joined the fleet, and more African and South American routes were added. SAT's first jetliner, a Convair CV-990A joined the fleet. Between 1968–1971, the fleet increased to 5 CV-990s, 5 DC-7Cs, 2 DC-6s, 7 DC-3s, and 4 Fokker F.27s . A Sabena swing-tail DC-6B was also acquired. SAT was the first charter company to offer nonstop Scandinavia-Canary Islands service in 1971. Transatlantic DC-7C charters were inaugurated in 1972. SPANTAX became the world's largest operator of the CV-990 after putting another 8 CV-990As into service between 1973 and 1977. By early 1988, the carrier was in deep financial trouble. A Kuwaiti investment firm indicated interest in providing cash aid; however, the Spanish government opposed the initiative. Unable to continue, Spantax ceased operations and declared bankruptcy at the end of March, claiming debts in excess of $80 million and leaving travel agents to deal with nearly 20,000 travelers who had already booked tickets. In the photo above is Spantax DC-7C, EC-BDM, at Gatwick airport on July 23, 1966. (RuthAS)

53. An Airlift International DC-7CF, N296, taxiing to takeoff position at San Francisco in Jan. 1967. Airlift International commenced operations on November 29, 1963, after previously operating under the name Riddle Airlines. With an extensive network of scheduled all-cargo services, the carrier was a major US passenger/cargo charter and military contract operator. In 1964, its fleet comprised 31 aircraft, mostly Curtiss C-46As and Douglas DC-7CFs, with a number of Lockheed L-1049H Super Connies. In July 1966 Airlift purchased Slick Airways, and its assets, which included Slick's Douglas DC-7CFs, increasing Airlift's DC-7CF fleet to 11. By 1969 all its DC-7CFs were retired. The US recession and cutback in Vietnam airlift operations brought a downturn for the carrier in 1970. In 1979, charter bookings shot up by 73.7% to 66,000 while cargo was down by 0.9% to 392.23 million. In 1981 Airlift filed for chapter 11 bankruptcy after having no physical assets, and no cash in hand. It continued to operate throughout the 1980s, but the writing was on the wall. Unable to be saved, Airlift ceased operations in 1993. (Bill Larkins)

54. Schreiner Airways was a Dutch-based charter, passenger, and cargo airline, founded in 1945 to represent aircraft manufacturers in the marketing of their equipment. Later, lightplane charter flights were undertaken for the Ministry of Defense. Schreiner Airways, in 1964/65, entered the charter market in competition with Martin Air Charters. The carrier utilized 3 DC-7 and a leased Fokker F.27 for regional flights. The fleet underwent expansion over the course of the following 2 years with the addition of 2 F.27s and 1 DC-7. After encountering significant challenges in the nonsked sector, Schreiner Airways, experienced a brief period of insolvency in 1967. However, it was subsequently revived as a subsidiary of the Scheiner Aviation Group, with a minority ownership stake held by KLM. That year, Schreiner removed the DC-7s and F.27s from its fleet. With no large airliners left in its fleet, Schreiner began to concentrate on operating contract services with small fixed-wing planes and helicopters. Its 3 DC-7Cs were operated on intercontinental charter flights between July 1965 and June 1968. Schreiner ceased operations in 2005. This photo is of DC-7C, PH-SAO. LSD in March 1966 from Pan Aero Corp., and RET in Nov. 1967. (RuthAS)

55. Caledonian Airways, was established on April 27, 1961 by John de la Haye and Adam Thomson, at Prestwick, Scotland, as an inclusive-tour carrier. The first charter flight was made from Barbados to London on November 29, with a Douglas DC-7C 'Star O' Robbie Burns,' leased from Sabena. Tragically, the original DC-7C with 10 crew and 101 passengers crashed on a Mozambique–Luxembourg charter flight on March 4, 1962, killing everybody on board. In April 1962, Another Sabena DC-7C was leased. Four more DC-7Cs and 3 Bristol Britannia 312s joined the fleet, opening further European and Mediterranean tourist destinations that year and the next. The Summer of 1965 saw 28 weekly transatlantic Britannia 312 and DC-7C flights. Travel to Africa and the Hadj retreat to Jeddah by air were also undertaken. In 1965, 3 of the 4 DC-7Cs were replaced by 3 Britannia 314s. Two Boeing 707-320Cs, the company's first jetliners, were ordered in November. After the final DC-7C was sold in the spring of 1967, the first Boeing 707-339C, 'Flagship Bonnie Scotland,' was accepted on July 13 and chartered to 'The Flying Tiger Line' for a year. 'The County of Ayr' launched Caledonian's jetliner service with a charter flight to North America on January 16, 1968. The bigger British United Airways was bought by Caledonian Airways on November 30, and the amalgamated carrier was named 'Caledonian/BUA' in 1970. The DC-7C in the photo was LSD to Caledonian April 5, 1962, and RET to SABENA in Oct. 1965. Later, STD in Brussels. Caledonian BT it 3 months later in Jan. 1966, and in March 1966, it sold it to German airline Sudflug. (Ken Fielding)

56. Flying Enterprise was a Danish Charter operator established in Copenhagen in May 1959 by two former SAS employees. Operations were started in January 1960 with one Argonaut. In 1961, the company became a subsidiary of the Heimdal A/S Shipping Company. It operated 5 Argonauts, but they were retired after the carrier received 4 104-seat DC-7s. The airline faced financial difficulties in 1964 and ceased operations in March 1965. The fleet of DC-7s was transferred to Conair, the airline that took over Flying Enterprise. Moreover, Flying Enterprise's Argonaut fleet was broken up at Copenhagen Airport, Kastrup. This DC-7, seen here at Malaga, southern Spain on Aug. 23, 1964, was sub-leased to Flying Enterprise by Sveaflyg on Dec. 21, 1963. In Feb. 1965 it was transferred to Osterman Air Charter. (Ken Fielding)

57. First created in 1950 as Nordisk Aerotransport, Transair Sweden changed its name in 1951. Early activities were newspaper delivery flights. Charter and inclusive-tour operations began with DC-3s in 1953. The first Palma de Mallorca charters started in 1954 with a technical stop at Hanover. Stockholm flights to Germany and southern Europe resumed in 1955–1956. In 1957, it began flying replacement services for SAS. Passenger and freight operations continued through 1958. On February 1, 1961, Transair offered an exciting new service, a 31-day excursion to McMurdo Sound, Antarctica. Northwest Airlines sold 2 DC-7Cs to Transair in 1965. Transair bought another 9 more DC-7s from Eastern Air Lines in 1966. However, technical issues with the turbo-compound engines made the new planes unpopular. Transair started joint operations with Scanair, a charter subsidiary of SAS, in 1968. The company moved to its new base at Sturup Airport near Malmo in 1972. In 1974, SAS terminated its agreement with Transair but was found guilty of breach of contract and fined $4.6 million. Rather than pay the huge settlement, SAS purchased Transair in 1975 and operated it as a subsidiary of Scanair. Business declined in 1980 due to the global financial crisis. Transair shut its doors in 1981. This DC-7B was bought by Transair from California Airmotive on Aug. 10, 1965. In Aug. 1968, it was WFU and STD at Malmo before being B/U in 1969. (Lars Söderström)

58. Transmeridian Air Cargo (TMAC), formerly known as Trans Meridian (London) until 1970, was founded in 1962 to provide international charter flights. It had two DC-7CFs in 1966, one of which was G-ATMF, photographed here at Manchester on June 25, 1966. TMAC bought this cargo '7CF' from US company Liberty Air. A year later, the airline bought 2 more DC-7CFs. Trans World Leasing bought TMAC in the end of 1967 and turned it into a pure freight operation, renaming it Transmeridian Air Cargo (TAC). The first of 8 CL-44s, which later replaced the DC-7CFs, arrived in 1968. In 1975, a base in Hong Kong was established for freight flights to Australia. TAC was purchased by the Trafalgar House Group in 1977, and on August 15, 1979, it was amalgamated with IAS Cargo Airlines to become British Cargo Airlines, which subsequently became HeavyLift Cargo. (Ken Fielding)

59. Miami-based All-American Airways was founded as a large irregular carrier in 1948, to provide charter and tour flights to North American locations, but, its main source of income was the US military. In 1964, the company, now with 8 Douglas DC- 6A/Bs and DC-7Cs, received CAB authority to operate a transatlantic route for the next 5 years. In April 1966, Saturn became one of 10 irregulars certified as supplemental carriers by the CAB. In 1967, Saturn bought 2 DC-8-61CFs, while its fleet also included 10 DC-7CFs and 3 DC-6As. In 1971, In 1971, Saturn became America's fifth largest supplemental carrier. It incorporated the L-100-30 Hercules as the workhorse of its cargo operations in Europe and throughout the Caribbean, often on behalf of the US military. Saturn was busy throughout the early 1970s. In 1973, it acquired 13 additional including 9 Lockheed L-188Fs. Then, in 1974, merger negotiations were undertaken with Trans-International Airlines (TIA), owned by Transamerica Corporation. The CAB approved the TIA amalgamation in the spring of 1976. On November 30, the purchase was officially completed and the company was merged into Trans-International Airlines. (Ken Fielding)

60. Sudflug was formed as private air carrier on Nov. 24, 1952 by the former German fighter pilot and self-made entrepreneur Rul Bückle. In its first years of operation it operated ad hoc charter for cargo transport and business flights with Cessna T-50s. Sudflug got its first De Havilland DH.114 Heron in 1955, which was used on Lufthansa scheduled flights from Stuttgart to Frankfurt am Main and Nuremberg. The carrier bought another Heron in 1956. Scheduled operations for Lufthansa ceased in autumn 1963. Sudflug also carried out charter work, specializing in executive flights. It also offered surveying, mapping, and geophysical survey flights in Africa. Sightseeing and joyriding is also undertaken. In 1964, the carrier purchased an ex-KLM DC-7CF. Sudflug received CAB foreign carrier air permission to operate group charter and inclusive tours to the USA, in 1965. The Herons were used for executive charter in Europe, and DC-7CFs for long range operations. In 1967, now known as Sudflug International, the carrier bought 2 DC-9s and 2 used Douglas DC-8s from Swissair. That caused it financial problems. Also, due to delays Swissair could not hand over the 2 DC-8s to Sudflug, until the fall of 1967. As a result, Sudflug had to lease two aircraft for the 1967 summer season to be able to fulfill the charter contracts they had already concluded. In 1967, the CAB reversed its decision and declined to authorize Sudflug's application for license to operate IT between Germany and the US. Lufthansa, the German national airline smelled blood, In a hostile takeover, Lufthansa took over the company on January 2, 1968 at a price of only 8 million DM. The DC-7 in the photo is D-ABAS, bought from KLM on Nov. 16, 1964. (Ken Fielding)

61. This DC-7B is in the colors of the 'Rank Organization' a British entertainment conglomerate, which is now defunct. The liner was leased by IBM in Sep. 1965, and somehow made its way to the Rank company, which possibly subleased it from IBM. The DC-7B was later leased by the band, 'Herman's Hermits' in May 1968. (Ken Fielding)

62. J. Martin Schroder and John Block created Martin's Air Charter (MAC) on May 22, 1958, with one aircraft, a de Havilland Dove ten seater. MAC offered leisure flights in its early years. On May 24, 1959, MAC began flying tulip bulbs to Berlin, using a Vickers Viking chartered from the British business Overseas Aviation. DC-3s were acquired in March and July 1960, and Amsterdam-Jersey and Mallorca IT flights were conducted for the tour operator Centouri. In 1964, MAC took up a large number of airline contracts from KLM, which purchased a 25% stake in MAC. Later that year, 2 DC-7Cs were added to the fleet and began charter operations. In 1966, the name was changed to Martinair Holland. Another DC-7C was acquired on February 23 . Furthermore, a MAC DC-7C became the first charter carrier airliner to fly the Polar route to Tokyo on March 24, 1966. One DC-7C was sold in October 1968, and another in April 1969. All MAC DC-7Cs were retired during the early 1970s. Today Martinair is still in operation, 65 years since its founding that is a great feat in the era of deregulation, and cut throat competition. The DC-7C above in PH-MAK, originally delivered to Alitalia on June 25, 1958. Sold to MAC on March 16, 1966. MAC, in turn, sold it to Liberty Air on March 1, 1967. (Ken Fielding)

63. SAS was formed on July 31, 1946. Following delivery of the 10 DC-4s ordered in 1943, the first SAS transatlantic New York service was inaugurated from Stockholm and Copenhagen on September 17, 1946 via Prestwick and Gander. In 1955 the first DC-7s were ordered by the company. The first DC-7C was delivered on August 16, 1956, and was named 'Brage Viking'; it was followed by the 'Guttorm Viking' on September 25. In 1957, orders were placed SAS' first jetliner, the French Sud SE-210 Caravelle. In March 1968 the last of 17 DC7Cs was retired by SAS. Of the 17 in the SAS fleet, 15 were DC-7Cs, and 2 were DC-7CFs. There were two incidents with the DC-7Cs. 1) 'SE-CCF' was struck at Khartoum airport on April 28, 1961, by an Ethiopian Airlines DC-6B, REG: ET-T-27, ran out of control when taxiing from one part of the apron to another. It hit 'SE-CCF' whose landing gear collapsed and was followed by fire breaking out. No casualties among the 8 crew and 32 pass on board financial W/O. 2) SE-CCC crashed during the takeoff run from Tenerife on Feb. 8, 1965. The crash was due to the premature retraction of the landing gear before the aircraft was airborne. No casualties among the 7 crew, and 84 passengers onboard. The DC-7C in the photo is LN-MOD, DEL: Sep. 25, 1956, c/n 44930, named 'Guttorm Viking'. It flew only with SAS, In April 1968, it was B/U at Kastrup-Copenhagen. (Ken Fielding)

64. This seemingly run-of-the-mill photo of a DC-7, N904ME, at London Airport on March 3, 1964, was operated by an insignificant company, Interpublic Inc. However, this photo has more to it than one would think. This is not just any DC-7C; this particular DC-7C was not only the last Douglas piston-engine airliner of the DC-6/7 family series built, it was also the last large piston-engine airliner built. When this DC-7C left the production line, it signified the end of the piston engine era, making this photo historic. Pyramid Oil Company was the propliner's last operator, acquiring it on Jan. 29, 1974. Its last flight was on Feb. 21, 1976. There were rumors this important DC-7C was to be restored in 2020; however, nothing has come of it. (Ken Fielding)

PHOTO CREDITS

9. Lars Söderström (https://commons.wikimedia.org/wiki/File:SAS_Douglas_DC-7_Soderstrom.jpg), „SAS Douglas DC-7 Soderstrom“, Cropped, enhanced color, brightness, tonal width, and sharpness by Theodore Gianna, https://creativecommons.org/licenses/by-sa/3.0/legalcode.

10. ETH-Bibliothek Zürich, Bildarchiv/Stiftung Luftbild Schweiz / Fotograf: Friedli, Werner / LBS_H1-028090 / CC BY-SA 4.0, Cropped, colorized, enhanced color, brightness, tonal width, and sharpness by Theodore Gianna.

11. ETH-Bibliothek Zürich, Bildarchiv/Stiftung Luftbild Schweiz / Fotograf: Swissair / LBS_SR04-037117 / CC BY-SA 4.0. enhanced color, brightness, tonal width, and sharpness by Theodore Gianna.

19. RuthAS (https://commons.wikimedia.org/wiki/File:BOAC_DC-7C_Taking-off_from_Manchester.jpg), „BOAC DC-7C Taking-off from Manchester“, colorized, enhanced color, brightness, tonal width, and sharpness by Theodore Gianna, https://creativecommons.org/licenses/by/3.0/legalcode.

20. Lars Söderström (https://commons.wikimedia.org/wiki/File:SAS_DC-7C_LN-MOE.jpg), „SAS DC-7C LN-MOE“, Enhanced color, brightness, tonal width, and sharpness by Theodore Gianna, https://creativecommons.org/licenses/by-sa/3.0/legalcode.

21. Lars Söderström (https://commons.wikimedia.org/wiki/File:SAS_DC-7C_LN-MOE.jpg), „SAS DC-7C LN-MOE“, Enhanced color, brightness, tonal width, and sharpness by Theodore Gianna, https://creativecommons.org/licenses/by-sa/3.0/legalcode.

26. Bill Larkins (https://commons.wikimedia.org/wiki/File:Douglas_DC-7F_N344AA_(6579626565).jpg), „Douglas DC-7F N344AA (6579626565)“, Enhanced color, brightness, tonal width, and sharpness by Theodore Gianna, https://creativecommons.org/licenses/by-sa/2.0/legalcode.

30. Nordiska museet (Nordic Museum) (https://commons.wikimedia.org/wiki/File:SAS_DC_7_i_luften_Färg_-_Nordiska_museet_-_NMAx.0011105.tif), Cropped, enhanced color, brightness, tonal width, and sharpness by Theodore Gianna, https://creativecommons.org/licenses/by-sa/4.0/legalcode

31. Maurice Luyten (https://commons.wikimedia.org/wiki/File:1958_SABENA_Fifties_Maurice_Luyten.jpg), Cropped, enhanced color, brightness, tonal width, and sharpness by Theodore Gianna, https://creativecommons.org/licenses/by/4.0/legalcode.

32. ETH-Bibliothek Zürich, Bildarchiv/Stiftung Luftbild Schweiz / Fotograf: Swissair / LBS_SR03-08392 / CC BY-SA 4.0. Changes: cropped, colorized, enhanced color, brightness, tonal width, and sharpness by Theodore Gianna.

37. Douglas DC-7 N6301C UAL, Bill Larkins, August 24, 2010, https://www.flickr.com/photos/34076827@N00/4922121917/in/photolist-8uXant-8uYLHh-8uXamZ. Colorized, enhanced color, brightness, tonal width, and sharpness by Theodore Gianna. CC BY-SA 2.0.

38. Nordiska museet (Nordic Museum) (https://commons.wikimedia.org/wiki/File:SAS_DC_7_i_luften_färg_-_Nordiska_museet_-_NMAx.0010990.tif), Cropped, enhanced color, brightness, tonal width, and sharpness by Theodore Gianna, https://creativecommons.org/licenses/by-sa/4.0/legalcode.

39. ETH-Bibliothek Zürich, Bildarchiv / Fotograf: Bärtschi, Hans-Peter / SIK_03-097394 / CC BY-SA 4.0..

41. ETH-Bibliothek Zürich, Bildarchiv / Fotograf: Furrer, Bernhard / SIK_02-07-0049 / CC BY-SA 4.0. Colorized, cropped, enhanced color, brightness, tonal width, and sharpness by Theodore Gianna.

43. RuthAS (https://commons.wikimedia.org/wiki/File:Douglas_DC-7CF_EI-ATT_Turas_LPL_06.69_edited-2.jpg), „Douglas DC-7CF EI-ATT Turas LPL 06.69 edited-2“, Enhanced color, brightness, tonal width, and sharpness by Theodore Gianna, https://creativecommons.org/licenses/by/3.0/legalcode.

44. Lars Söderström (https://commons.wikimedia.org/wiki/File:Internord_DC-7_SE-CNF.jpg), „Internord DC-7 SE-CNF“, Enhanced color, brightness, tonal width, and sharpness by Theodore Gianna, https://creativecommons.org/licenses/by-sa/3.0/legalcode.

45. RuthAS (https://commons.wikimedia.org/wiki/File:Douglas_DC-7B_N4882C_Vance_LGB_06.02.71_edited-2.jpg), „Douglas DC-7B N4882C Vance LGB 06.02.71 edited-2“, Cropped, enhanced color, brightness, tonal width, and sharpness by Theodore Gianna, https://creativecommons.org/licenses/by/3.0/legalcode.

46. Ken Fielding/https://www.flickr.com/photos/kenfielding (https://commons.wikimedia.org/wiki/File:EC-BEO_DC-7C_TAE_PMI_17SEP67_(5575027341).jpg), „EC-BEO DC-7C TAE PMI 17SEP67 (5575027341)“, Cropped, enhanced color, brightness, tonal width, and sharpness by Theodore Gianna, https://creativecommons.org/licenses/by-sa/3.0/legalcode.

47. Ken Fielding/https://www.flickr.com/photos/kenfielding (https://commons.wikimedia.org/wiki/File:G-AOII_DC-7C(F)_BOAC_Cargo_LHR_03APR64_(5575028613).jpg), „G-AOII DC-7C(F) BOAC Cargo LHR 03APR64 (5575028613)“, Cropped, enhanced color, brightness, tonal width, and sharpness by Theodore Gianna, https://creativecommons.org/licenses/by-sa/3.0/legalcode.

48. Ken Fielding/https://www.flickr.com/photos/kenfielding (https://commons.wikimedia.org/wiki/File:EC-AVP_DC-7_TASsa_(stored)_PMI_11SEP65_(5575026393).jpg), „EC-AVP DC-7 TASsa (stored) PMI 11SEP65 (5575026393)“, Cropped enhanced color, brightness, tonal width, and sharpness by Theodore Gianna, https://creativecommons.org/licenses/by-sa/3.0/legalcode.

49. Lars Söderström (https://commons.wikimedia.org/wiki/File:Conair_Douglas_DC-7_Söderström-2.jpg), „Conair Douglas DC-7 Söderström-2“, Cropped, enhanced color, brightness, tonal width, and sharpness by Theodore Gianna, https://creativecommons.org/licenses/by-sa/3.0/legalcode.

51. Lars Söderström (https://commons.wikimedia.org/wiki/File:Osterman_Air_Charter_Douglas_DC-7_Soderstrom-2.jpg), „Osterman Air Charter Douglas DC-7 Soderstrom-2“, Cropped, enhanced color, brightness, tonal width, and sharpness by Theodore Gianna, https://creativecommons.org/licenses/by-sa/3.0/legalcode.

52. RuthAS (https://commons.wikimedia.org/wiki/File:Douglas_DC-7C_EC-BDM_Spantax_LGW_23.07.66_edited-3.jpg), „Douglas DC-7C EC-BDM Spantax LGW 23.07.66 edited-3“, Cropped, enhanced color, brightness, tonal width, and sharpness by Theodore Gianna, https://creativecommons.org/licenses/by/3.0/legalcode.

53. Bill Larkins (https://commons.wikimedia.org/wiki/File:Douglas_DC-7C_N296_(7859000496).jpg), „Douglas DC-7C N296 (7859000496)“, Enhanced color, brightness, tonal width, and sharpness by Theodore Gianna, https://creativecommons.org/licenses/by-sa/2.0/legalcode.

54. RuthAS (https://commons.wikimedia.org/wiki/File:Douglas_DC-7C_PH-SAO_Schreiner_AMS_12.03.67_edited-2.jpg), „Douglas DC-7C PH-SAO Schreiner AMS 12.03.67 edited-2“, Cropped, enhanced color, brightness, tonal width, and sharpness by Theodore Gianna, https://creativecommons.org/licenses/by/3.0/legalcode.

55. Ken Fielding/https://www.flickr.com/photos/kenfielding (https://commons.wikimedia.org/wiki/File:Caledonian_Airways_Douglas_DC-7C_(G-ARYE)_at_the_Biggin_Hill_International_Air_Fair.jpg), „Caledonian Airways Douglas DC-7C (G-ARYE) at the Biggin Hill International Air Fair“, Cropped, enhanced color, brightness, tonal width, and sharpness by Theodore Gianna, https://creativecommons.org/licenses/by-sa/3.0/legalcode.

56. Ken Fielding/https://www.flickr.com/photos/kenfielding (https://commons.wikimedia.org/wiki/File:OY-DMR_DC-7_Flying_Enterprise_AGP_23AUG64_(5575031533).jpg), „OY-DMR DC-7. Flying Enterprise AGP 23AUG64 (5575031533)“, Cropped, enhanced color, brightness, tonal width, and sharpness by Theodore Gianna, https://creativecommons.org/licenses/by-sa/3.0/legalcode.

57. Lars Söderström (https://commons.wikimedia.org/wiki/File:Transair_Sweden_Douglas_DC-7_Soderstrom.jpg), „Transair Sweden Douglas DC-7 Soderstrom“, Cropped, enhanced color, brightness, tonal width, and sharpness by Theodore Gianna, https://creativecommons.org/licenses/by-sa/3.0/legalcode.

58. Ken Fielding/https://www.flickr.com/photos/kenfielding (https://commons.wikimedia.org/wiki/File:G-ATMF_DC-7C_F_Trans_Meridian_Aws_MAN_25JUN66_(5575616132).jpg), „G-ATMF DC-7C F Trans Meridian Aws MAN 25JUN66 (5575616132)“, Cropped, enhanced color, brightness, tonal width, and sharpness by Theodore Gianna, https://creativecommons.org/licenses/by-sa/3.0/legalcode.

59. Ken Fielding/https://www.flickr.com/photos/kenfielding (https://commons.wikimedia.org/wiki/File:N90773_DC-7C_Saturn_Aws_MAN_25JUL65_(5575617464).jpg), „N90773 DC-7C Saturn Aws MAN 25JUL65 (5575617464)“, Cropped, enhanced color, brightness, tonal width, and sharpness by Theodore Gianna, https://creativecommons.org/licenses/by-sa/3.0/legalcode.

60. Ken Fielding/https://www.flickr.com/photos/kenfielding (https://commons.wikimedia.org/wiki/File:D-ABAS_DC-7C_Sudflug_LPL_03MAR65_(5575023897).jpg), „D-ABAS DC-7C Sudflug LPL 03MAR65 (5575023897)“, Cropped, enhanced color, brightness, tonal width, and sharpness by Theodore Gianna, https://creativecommons.org/licenses/by-sa/3.0/legalcode.

61. Ken Fielding/https://www.flickr.com/photos/kenfielding (https://commons.wikimedia.org/wiki/File:N774R_DC-7B_Rank_Organisation_MAN_25JUN66_(5575617010).jpg), „N774R DC-7B Rank Organisation MAN 25JUN66 (5575617010)“, Cropped, enhanced color, brightness, tonal width, and sharpness by Theodore Gianna, https://creativecommons.org/licenses/by-sa/3.0/legalcode.

62. Ken Fielding/https://www.flickr.com/photos/kenfielding (https://commons.wikimedia.org/wiki/File:PH-MAK_DC-7C_Martins_Air_Chtr_LPL_23JUL66_(5575032713).jpg), „PH-MAK DC-7C Martins Air Chtr LPL 23JUL66 (5575032713)“, Cropped enhanced color, brightness, tonal width, and sharpness by Theodore Gianna, https://creativecommons.org/licenses/by-sa/3.0/legalcode.

63. Nordiska museet (Nordic Museum) (https://commons.wikimedia.org/wiki/File:SAS_DC_7_i_luften_Färg_-_Nordiska_museet_-_NMAx.0011104.tif), Cropped, enhanced color, brightness, tonal width, and sharpness by Theodore Gianna, https://creativecommons.org/licenses/by-sa/4.0/legalcode.

64. Ken Fielding/https://www.flickr.com/photos/kenfielding (https://commons.wikimedia.org/wiki/File:N904ME_DC-7C_Starflights_Inc_LHR_03MAR64_(5575617272).jpg), „N904ME DC-7C Starflights Inc LHR 03MAR64 (5575617272)“, Cropped, enhanced color, brightness, tonal width, and sharpness by Theodore Gianna, https://creativecommons.org/licenses/by-sa/3.0/legalcode.

FRONT COVER PHOTO: Nordiska museet (Nordic Museum) (https://commons.wikimedia.org/wiki/File:SAS_DC_7_i_luften_färg_-_Nordiska_museet_-_NMAx.0011025.tif), Cropped, enhanced color, brightness, tonal width, and sharpness by Theodore Gianna, https://creativecommons.org/licenses/by-sa/4.0/legalcode.

BACK COVER PHOTO: Lars Söderström (https://commons.wikimedia.org/wiki/File:Conair_DC-7_OY-DMT.jpg), „Conair DC-7 OY-DMT“, Enhanced color, brightness, tonal width, and sharpness by Theodore Gianna, https://creativecommons.org/licenses/by-sa/3.0/legalcode.

Made in the USA
Las Vegas, NV
03 November 2026

57423050R00048